Anthropomorphic Phantoms in Image Quality and Patient Dose Optimization

A EUTEMPE Network book

IPEM–IOP Series in Physics and Engineering in Medicine and Biolog

About the Series

Series in Physics and Engineering in Medicine and Biology will allow IPEM to enhance its mission to 'advance physics and engineering applied to medicine and biology for the public good.'

Focusing on key areas including, but not limited to:

- clinical engineering
- diagnostic radiology
- informatics and computing
- magnetic resonance imaging
- nuclear medicine
- physiological measurement
- radiation protection
- radiotherapy
- rehabilitation engineering
- ultrasound and non-ionising radiation.

A number of IPEM–IOP titles are published as part of the EUTEMPE Network Series for Medical Physics Experts.

Anthropomorphic Phantoms in Image Quality and Patient Dose Optimization

A EUTEMPE Network book

Kristina Bliznakova, Ivan Buliev and Zhivko Bliznakov
Technical University of Varna, Bulgaria

IOP Publishing, Bristol, UK

ISBN 978-0-7503-1323-0 (ebook)
ISBN 978-0-7503-1324-7 (print)
ISBN 978-0-7503-1325-4 (mobi)

DOI 10.1088/2053-2563/aae197

Version: 20181201

IOP Expanding Physics
ISSN 2053-2563 (online)
ISSN 2054-7315 (print)

British Library Cataloguing-in-Publication Data: A catalogue record for this book is available from the British Library.

Published by IOP Publishing, wholly owned by The Institute of Physics, London

IOP Publishing, Temple Circus, Temple Way, Bristol, BS1 6HG, UK

US Office: IOP Publishing, Inc., 190 North Independence Mall West, Suite 601, Philadelphia, PA 19106, USA

Contents

Preface

Dear Learner,

This book is dedicated to the module on Anthropomorphic phantoms for optimisation of dose and image quality in radiology from the European Network for Training and Education of Medical Physics Experts—EUTEMPE (www.eutempe-net.eu).

The present module aims to broaden the knowledge of medical physics experts and biomedical engineers, and develop practical skills in the design, creation, and use of anthropomorphic phantoms—physical objects or computer models which replace tissues and organs in real or virtual experiments. Using anthropomorphic phantoms is considered a valuable complementary instrument for research, hypothesis testing, optimisation of different diagnostic or therapeutic procedures, etc.

The module is prepared and delivered by biomedical engineers and there is nothing strange in that. Biomedical engineers manage, design, prototype, and manufacture new diagnostic and therapeutic equipment. Along with this, together with medical physicists and other specialists, they usually need to perform a large number of experiments and tests. Unfortunately, when humans are targets, and especially when hazardous techniques (such as applying ionising radiation) are used, this is no longer feasible. Working with objects which roughly or precisely resemble anatomical structures, organs, and the whole human body has proven to be an excellent alternative. The manufacture of such objects can even be avoided or at least significantly optimised with the help of realistic computer simulations. The suggested module content is strongly related to the experience of the involved module leader and assistants.

We hope that through reading the book and attending this module of the EUTEMPE course, you will professionally benefit and enjoy it at the same time.

Welcome on board!

Author biographies

Kristina Bliznakova

Kristina Bliznakova received MSc and PhD degrees in biomedical engineering from the University of Patras, Greece. She has been a Marie Curie fellow at the Technical University of Varna, Bulgaria, and she has been awarded the MSCA 2017 first prize as the most promising scientist. Currently, she is an Associate Professor at the Technical University of Varna, teaching computer programming and software development topics. During the last two decades, she has been actively conducting research activities, leading also several national and international research projects.

Ivan Buliev

Ivan Buliev received his diploma in electrical engineering from the Technical University of Varna, Bulgaria, and MSc and PhD degrees in biomedical engineering from the University of Patras, Greece. Currently, he is an Associate Professor at the Technical University of Varna, teaching embedded systems, industrial electronics, medical electronics, biomedical signal processing and medical imaging.

Zhivko Bliznakov

Zhivko Bliznakov received his diploma in electrical engineering from the Technical University of Varna, Bulgaria, and MSc and PhD degrees in biomedical engineering from the University of Patras, Greece. During the last two decades, he has been actively involved in the field of biomedical engineering, performing research and educational activities. He has been also managing a high number of national and European projects.

Chapter 1

Implementation of the module on anthropomorphic phantoms

1.1 Introduction

Anthropomorphic phantoms in radiology permit safe unlimited exposure. They are used in different tasks such as image quality assessment, patient dose optimisation, clinical commissioning, and pre-clinical trials in health technology assessments.

The main aim of this book is to acquaint readers with the nature and the role of physical and virtual anthropomorphic phantoms in performing virtual (phantom-based) clinical trials using existing and new diagnostic and interventional radiology technologies. Readers will be introduced to the different existing types of anthropomorphic phantoms used in clinical trials, and will be encouraged to develop skills for the design and evaluation of anthropomorphic phantoms, as well as to design, manage, implement, and evaluate virtual clinical studies with such phantoms, and to discuss and interpret the results of the virtual studies.

The relevant EUTEMPE module, on the other hand (should you decide to attend it), uses a blended teaching format that includes preliminary readings, lectures, computer-based exercises, manufacturing of phantoms, visits to hospital for experimental work, and discussion sessions.

1.2 Content

The content of the module is organised in the following groups of topics:

- *Anthropomorphic phantoms*. Definition. Classification of phantoms. State-of-the-art review of anthropomorphic physical and computational phantoms. Design and composition of both physical and computational anthropomorphic phantoms.

doi:10.1088/2053-2563/aae197ch1

- *SW tools in x-ray imaging research.* Tools for image processing *MATLAB* (optional), *Octave*, and *ImageJ*, tools for phantom modelling and simulation of imaging techniques (*LUCMFRGen*, *BreastSimulator*, and *XRAYImagingSimulator*) and a tool for image reconstructions (*FDKR*). The last three applications are in-house developed software tools. Exercises include: design and construction of anthropomorphic phantoms and use of anthropomorphic phantoms in (a) optimising the parameters of an existing imaging modality, (b) design and evaluation of advanced x-ray imaging techniques, (c) development and testing of image reconstruction techniques, (d) assessing the limitations of an x-ray imaging modality, and (e) development and testing of image enhancement techniques.
- *Computational anthropomorphic phantoms as prototypes of physical objects.* Creation of anthropomorphic phantoms. Segmentation of tissues to be printed. Conversion of computational phantoms to stereolithography (STL) format for printing. Working with dedicated software for 3D printing; as part of this training the trainee will learn how to place the support structures, to print the model, and to remove the printed support structures.

The content of this book is organised as follows: chapter 1 is composed by all authors, chapters 2–5 and 7 were developed by Kristina Bliznakova, chapter 6 was prepared by Ivan Buliev, and chapter 8 was written by Zhivko Bliznakov. All three authors are biomedical engineers with PhDs in biomedical engineering and high-level expertise in the field of anthropomorphic phantoms and related applications. Chapter 1 introduces the module, while chapter 2 introduces anthropomorphic phantoms, their classification, and use. Chapter 3 deals with approaches for the creation of anthropomorphic phantoms: geometry and composition and the advantages of each of the chosen representations. Chapter 4 focuses on the practical design and creation of anthropomorphic phantoms and image formation. The specific case of the breast is considered in chapter 5. The reconstruction of projection images obtained in 3D imaging mode (computed tomography or tomosynthesis) is discussed in chapter 6. These three main chapters (4–6) provide training for the main tools to be used in carrying out virtual clinical studies, the concern of chapter 7. Finally, chapter 8 demonstrates the complete chain from the design of the computational phantom to the production of its physical prototype using 3D printing technologies.

1.3 Learning objectives

The learning objectives have been preliminarily identified and set. The Learners will be able to:

- Demonstrate *knowledge* of computational/physical anthropomorphic phantoms and their application in diagnostic and interventional radiology.
- Design and implement *new physical* and *virtual anthropomorphic* phantoms.
- Use software tools to *design anthropomorphic phantoms*.

- Run virtual clinical and research studies with anthropomorphic phantoms to *optimise or develop a clinical protocol.*
- Run virtual clinical and research studies with anthropomorphic phantoms to assess the *capabilities and limitations of an imaging modality.*
- Run virtual clinical and research studies with anthropomorphic phantoms to *optimise the parameters* of an existing imaging device.
- Run virtual clinical and research studies with anthropomorphic phantoms to develop and evaluate *image enhancement techniques.*
- Run virtual clinical and research studies with anthropomorphic phantoms to develop and evaluate *image reconstruction algorithms.*
- Simulate an *imaging experiment* prior to its real set-up.
- Run virtual research studies with anthropomorphic phantoms to perform *advanced scientific research* related to emerging x-ray imaging techniques.
- *Understand* when, why and how to use anthropomorphic phantoms in daily clinical practical and research diagnostic and interventional radiology work.

1.4 Teaching approach

For those who will register for the module, the training will be organised in two phases—a preliminary *online phase* (e-learning) and an essential *face-to-face session*, held about two months later at the Technical University of Varna, Bulgaria.

The first, online phase is provided through a Moodle platform. The main aim of this phase is to provoke the participants to read and acquire some preliminary information on the topics that will be discussed in the face-to-face session, and to become familiar with the software tools that are going to be used. Relevant information is presented in the form of presentations, videos, interactive applications, and references to optional articles. We will expect you to become familiar with a few in-house developed software tools (*XRAYImagingSimulator*, *BreastSimulator*, *LUCMFRGen*, and *FDKR*), which run on the Microsoft Windows OS. Text and video tutorials on how to use them are also provided and it will be very useful if you try to complete the examples described in the following sections. Some topics will be further discussed during the brainstorming sessions in Varna.

The second phase is the face-to-face session at the Technical University of Varna. For the different module topics we will first make an overview of the preliminary information, and then we will go into more detail. We will provide you with demonstrative examples and tutorials and, most importantly, we will provide the possibility to work with the tools and solve practical problems, such as the design and creation of your own anthropomorphic phantom, its application in various virtual studies such as optimising tomosynthesis, investigating dual energy or phase contrast imaging using anthropomorphic phantoms created by you, image reconstruction techniques, etc. An interesting part should be the real production of physical phantoms from computational models using our 3D printing facilities. We believe you will enjoy the session in Varna!

1.5 Assessment

Every educational activity normally ends with some sort of assessment. In this module the assessment will consist of an implementation of a work project on a case study from diagnostic and interventional radiology, combined with a short written exam. The work project will be assigned to you at the beginning of the face-to-face session and you will have to carry it out throughout the week. Examples of work projects are given in chapter 7. Establishment of working groups will be based on a procedure of random member selection. You will work in groups of two and you will have to conclude with a short (about 10 min) presentation shared fairly between the two participants. The written exam will consist of about 20 short questions (some of them will be multiple choice questions) and will have a duration of one hour. The final assessment mark will be calculated on the basis of 60% project mark (30% project development and 30% project presentation) and 40% written exam mark. The final score of the assessment is: < 50% fail, 50%–64% pass, 65%–79% merit, and 80%–100% distinction. The participants' level of English will not be taken into account in the evaluation.

1.6 Computer facilities

Learners may use laptops from the Technical University of Varna running a 64 bit Microsoft Windows OS. However, we encourage you to take your own portable computers to Varna. It should be more convenient for you to work on them and perform the computations. You can also copy all collected module materials directly to your own storage and have them ready for use in further studies.

All additional information and software applications which you may need to develop your exercises will be provided to you on-site. You do not need to have a solid background in computer programming. However, knowledge of *MATLAB* or *Octave* will be advantageous to grasp some concepts and work out some exercises in detail, specifically those related to the data processing prior to the 3D printing.

1.7 Supplementary material

Supplementary material in the form of video, interactive applications, scripts, and code can be found on the website of the course module. Relevant links are provided throughout the text.

Chapter 2

Introduction to anthropomorphic phantoms

2.1 Introduction

Anthropomorphic phantoms are real or virtual objects composed of tissue-equivalent materials that aim to provide a realistic and accurate representation of the anatomy and the properties of tissues, organs, and the whole body for use in real and virtual studies carried out in research and daily clinical tasks. Realistic and accurate presentation means to correctly mimic the outer shape and/or the internal structures of the modelled human tissue and organs. The properties of the phantom depend on the properties of the tissue substitutes used in the specific diagnostic application. In x-ray imaging, the tissue substitute is any material and thickness that simulates the x-ray attenuation and scattering of human anatomy.

In this chapter, attention is focussed on the computer and physical models of anthropomorphic phantoms dedicated to tasks in diagnostic radiology. The aim is to classify these phantoms according to their use, complexity, and parts, and to demonstrate some favourite applications in the field of x-ray imaging. An overview of the main components of both physical and computational phantoms is listed as well. In the case of computational techniques, modelling approaches are presented together with their advantages and disadvantages. At the end of the chapter test questions and quizzes are provided.

2.2 Anthropomorphic phantom classification

This section presents phantoms with respect to some classification schemes.

- *Phantoms classified with respect to their purpose.*

In general, anthropomorphic phantoms dedicated to tasks in diagnostic radiology may be classified in two basic categories: (a) for testing, evaluation and overall

doi:10.1088/2053-2563/aae197ch2

imaging assessment of complete (or parts) of existing (or new) x-ray imaging techniques and (b) for *dose assessment*; see figure 2.1.

Phantoms for imaging assessment are phantoms used (i) to assess the image quality of existing and newly developed x-ray imaging techniques, (ii) to test the limitations of x-ray imaging systems, for instance the achievable resolution of x-ray beam and detector systems, the amount of contrast needed to distinguish objects from one another, (iii) for research and development involving *in vitro* experimental validation prior to clinical application in humans, (iv) for optimisation of image quality and radiation exposure, (v) for evaluation of dose and scattered radiation, (vi) for development and optimisations of reconstruction algorithms, and (vii) as quality assurance (QA) tools for initial implementations of an imaging protocol and routine quality control (QC) for the imaging modality. These phantoms may be both computational and physical.

Phantoms for dose assessment are used to assess the dose from medical imaging procedures. These phantoms may also be computational and physical. It is clearly impractical and dangerous to place an actual human in the beam to take measurements, and perhaps even more impractical to place dosimeters inside a human to take measurements. This was the impetus for the first dosimetry phantoms, which were tanks of water or slabs of tissue-equivalent materials designed to hold dosimeters and to allow for measurements 'in tissue' without any unnecessary exposure to the people involved.

- *Phantoms classified with respect to their nature as either physical or computational.*

Computational phantoms are descriptions of human tissues/organs and bodies using computer tools, dedicated algorithms, and their storage. They can be quite sophisticated representations of part of or the complete human body in terms of

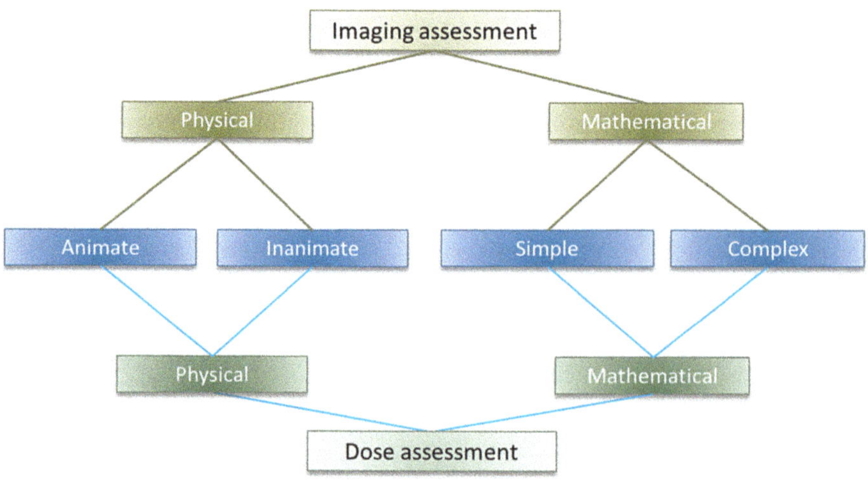

Figure 2.1. Classification of anthropomorphic phantoms by their purpose.

shape, size, and distribution of different organs. Alternatively, they might be quite simple, dedicated to depth dose assessment and evaluation of beam qualities, as in these cases they might be represented by simple geometric 3D primitives. The primary aim of these phantoms is to model the interaction of the x-ray beams with the biological tissues in order to evaluate the absorbed and the scattered dose, as well as image noise. These calculations can be achieved either by using Monte Carlo techniques or by implementation of simple analytical and semi-analytical approaches (such as the Beer–Lambert law).

Physical phantoms are objects manufactured using moulding or 3D printing techniques with materials having x-ray imaging characteristics close to these of human tissues. Physical phantoms might range from a single block of material (for instance a slab of PMMA to represent the breast or a water cylinder to represent part of the human body) to more sophisticated structures containing a variety of details.

- *Phantoms classified with respect to the inclusion of living entities.*

Animate phantoms constitute living entities, either human or animal, and can be classified in the phantom domain. These phantoms are frequently used in the evaluation of the performance of radiological systems and techniques: to verify the functional capabilities of every new radiological technique and to assess its performance relative to other accepted diagnostic techniques.

Inanimate phantoms possess real or simulated anatomic structures or details. These include for instance phantoms which contain skeletal details or mimic breast parenchyma. Cadaver sections also fall into this category. An example is shown in figure 2.2 for the case of breast cadaver. Inanimate phantoms can be used to assess imaging performance either subjectively or by physical means. For example, computed tomography (CT) phantoms, which simulate anatomical cross sections, permit consistent evaluation of CT number values and allow detailed noise analysis.

- *Phantoms classified by the complexity of their design and construction.*

A more detailed classification, taking into account the design and the construction of the phantoms, results in four groups, representatives of which are depicted in figure 2.3.

Specifically, with respect to this classification, phantoms may be grouped as follows:

- *Phantoms represented by simple shapes and content.* These phantoms reflect the basic absorption properties of the tissue of interest and at the same time are simple in form in order to be reproduced easily (for instance a physical version). Figures 2.3(a), (e) show examples of both physical and computational phantoms. The phantom shown in figure 2.3(a) is the GAMEX 156 phantom, which is a breast physical phantom dedicated to monitoring image quality and dose. The phantom shown in figure 2.3(e) is a computational paraffin phantom with included water spherical features (mimicking breast masses), dedicated to testing the visibility of lesions under different x-ray imaging techniques.

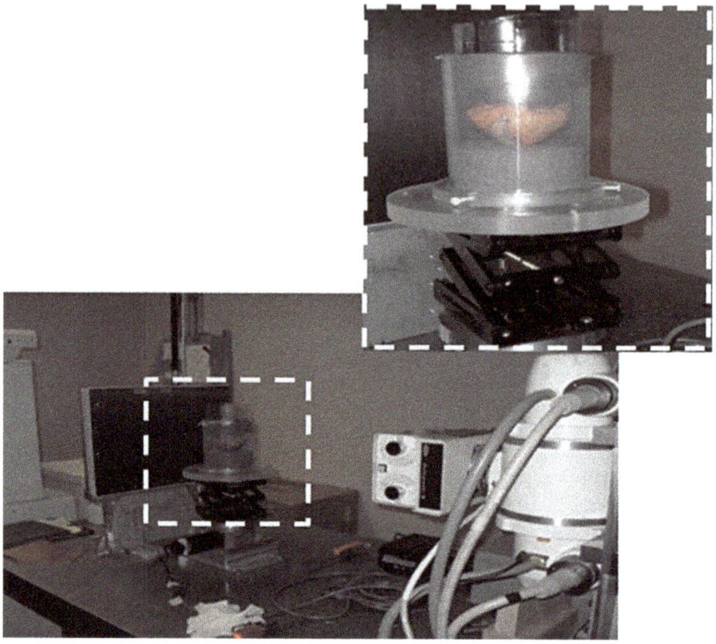

Figure 2.2. An example of inanimate phantoms: the use of breast cadaver in studies of radiation doses from cone-beam CT. (Reprinted from Yi *et al* 2011 © 2011 American Association of Physicists in Medicine.)

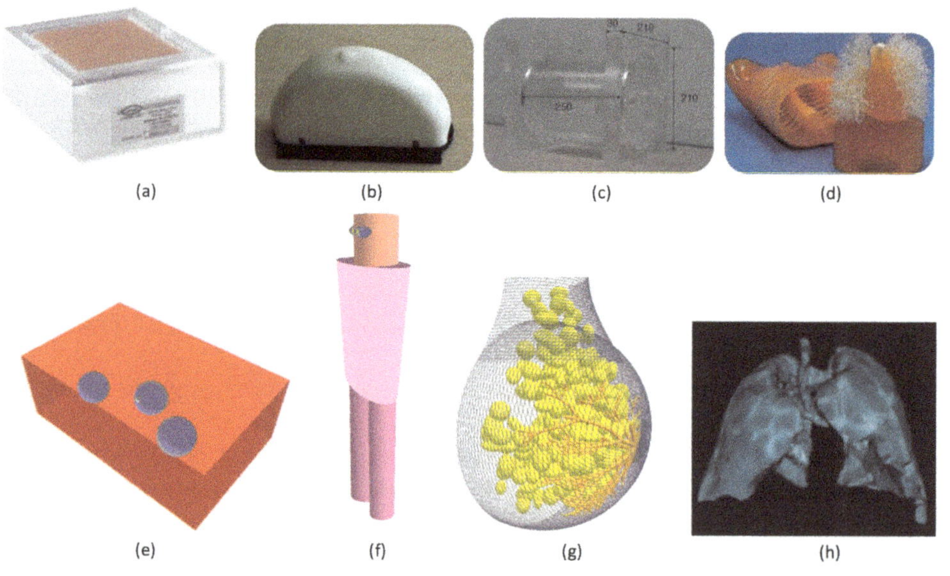

Figure 2.3. Classification of phantoms by design and construction: (a, e) simple in shape and content, (b, f) with anthropomorphic outward physical shape and inaccurate models of anatomical internal structures, (c, g) phantoms with simple exterior geometries containing detailed models of internal structures (panel (c) reproduced from Han *et al* 2008 © 2008 American Association of Physicists in Medicine), and (d, h) phantoms with detailed modelled internal structures (panel (h) reproduced from Busse *et al* 2013 © 2008 American Association of Physicists in Medicine). The phantoms shown in (a–d) are physical phantoms, while those in (e–h) are computational phantoms.

- *Phantoms which are anthropomorphic in their outward physical shape but do not necessarily contain anatomically accurate internal structures.* The phantoms, shown in figures 2.3(b), (f), are examples of this type of phantom. Figure 2.3(b) depicts the CIRS breast phantom 011A, which is a tissue-equivalent mammography QA phantom routinely used in the work of radiology departments. The phantom shown in figure 2.3(f) is a computational full-body phantom used to simulate the doses of operational workers during diagnostic and interventional procedures.
- *Phantoms with simple exterior geometries but containing detailed internal structures.* Figures 2.3(c), (g) show two examples of such phantoms: the breast phantom with an external shape formed by two 3D geometrical primitives, a semi-ellipsoid and hyperboloid, filled with accurately modelled breast structures (figure 2.3(g)), and a physical head/neck phantom implemented as a cylinder with a detailed representation of anatomy (figure 2.3(c)).
- *Phantoms with anatomically accurate and detailed exterior shape and internal structures.* The Lungman N1 multipurpose chest phantom (Kyoto Kagaku, Kyoto, Japan) (figure 2.3(d)) and the computational model of the lungs (figure 2.3(h)) are two excellent examples of this type of phantom. These are used to accurately simulate the procedure of interest.

An in-house developed software tool *Classification of Phantoms* will be available for the participants in the EUTEMPE course.

2.3 Anthropomorphic physical phantoms for x-ray imaging

The following two sections focus on the physical and computational anthropomorphic phantoms, their design, complexity, use, advantages, and limitations. Depending on the extent to which the human body is modelled, the phantoms may be full-body phantoms or phantoms that model parts of the body.

2.3.1 Full-body physical phantoms

Full-body phantoms are usually body phantoms with dimensions of an average-sized male or female patient, as well as children of different ages. These phantoms are mainly used to train medical students in how to use radiography equipment and produce high-quality radiographic images. Some, but not limited to these, full-body phantoms are the RANDO phantom, the Kyoto Kagaku CT phantom, and the CIRS ATOM® phantoms.

The RANDO® phantom (The Phantom Laboratory) is represented by two RANDO® models: the RANDO® woman and the RANDO® man. The RANDO® woman represents a 163 cm tall and 54 kg female, while the RANDO® man represents a 175 cm tall and 73.5 kg male. Both models, shown in figure 2.4, do not have arms or legs. The phantoms are constructed with a natural human skeleton (1.61 g cm^{-3}) which is cast inside soft tissue-simulating material. Lungs are moulded to fit the contours of the natural rib cage. The air spaces of the head, neck and stem bronchi are duplicated.

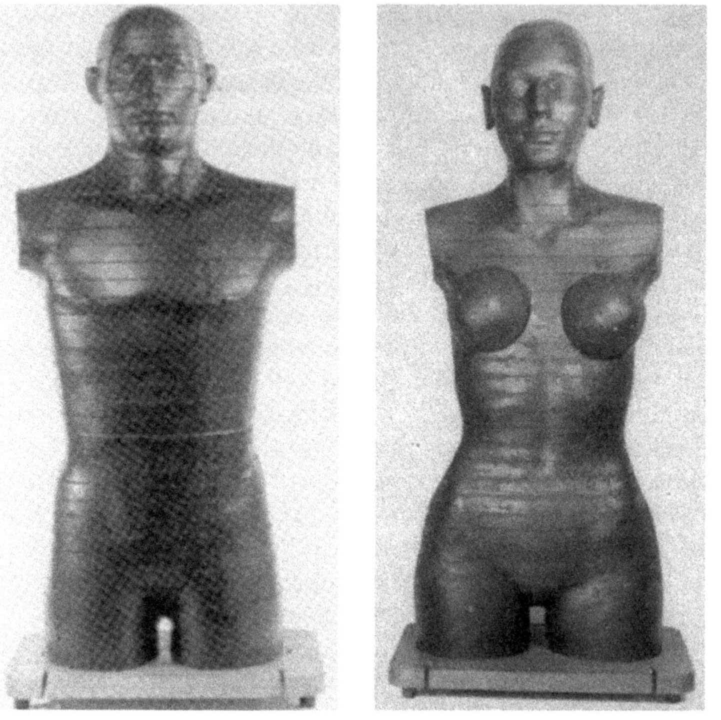

Figure 2.4. The RANDO® male (Struelens *et al* (2008)) and the RANDO® female (Abella *et al* 2010) [Reprinted with the permission of Elsevier] phantoms.

The phantom is composed of 33 slices with a thickness of 2.5 cm in order to insert films between slices.

Two tissue-simulating materials are used to construct the RANDO® phantom: the RANDO® soft tissue material (0.977 g cm^{-3}) and the RANDO® lung material. The RANDO® phantom's soft tissue is manufactured with a proprietary urethane formulation. The material has an effective atomic number and mass density which simulate muscle tissue with randomly distributed fat. The RANDO® phantom's lung material has the same effective atomic number as the soft tissue material with a density which simulates lungs in a median respiratory state. The moulded lungs are hand-shaped and fitted to naturally fill the rib cage. Breasts in sizes A through E are also available for both female and male phantoms. Each breast is moulded into 3 cm thick sections.

This phantom is mainly used in radiotherapy treatment planning. It enables detailed mapping of dose distribution, which is useful for teaching patient treatment evaluation and quality assurance of automated treatment planning systems and research. Due to the moulding process, small air bubbles are formed which may be evident in the x-ray images. These are small enough to not have a significant effect on the virtual clinical studies.

The Kyoto Kagaku CT phantom, shown in figure 2.5, is a unique, life size whole-body phantom dedicated mainly to CT imaging, which provides a variety of educational applications as well as visual evaluation in determining optimal scanning conditions.

Figure 2.5. Kyoto Kagaku life size whole-body phantom (www.kyotokagaku.com/products/detail03/ph-2.html).

The phantom can also be used for planar x-ray imaging. It is composed of a state-of-the-art synthetic skeleton, lungs, liver, mediastinum, and kidneys embedded in the original soft tissue substitute. There are no metal parts or liquid structure within the phantom.

In this *patient specific* phantom, 24 organs are modelled, which are anatomically correct and have appropriate Hounsfield units (HU). For soft tissue and organs a urethane-based resin (SZ-50) is used, for the synthetic bones and epoxy-based resin is used, while for joint attachments epoxy and urethane with carbon fibre is utilised. The screws are made from polycarbonate. Among the important applications of this phantom is the assessment of CT imaging performance.

CIRS ATOM® phantoms are a full line of anthropomorphic, cross-sectional dosimetry phantoms designed to investigate organ dose, whole-body effective dose, as well as verification of the delivery of therapeutic radiation doses. ATOM® is the only line of dosimetry phantoms to range in size from new-born to adult (figure 2.6).

Six ATOM models are available: new-born, 1-year-, 5-year-, and 10-year-old paediatric phantoms as well as adult male and female phantoms. Each phantom is sectional in design with traditional 25 mm thick sections. The sectional surfaces are extremely flat and smooth and do not require any special coatings or treatment. This results in minimal interfaces between the slabs when viewed in a scout or x-ray projection images. Tissue-equivalent epoxy resins are the main materials for manufacturing the different parts of the phantom. In addition, all bones are homogeneous and are formulated to represent age appropriate, average bone composition, which offers distinct advantages over natural skeletons and other types of simulated bone.

The Pediatric anthropomorphic phantom CIRS 715 represents a typical 5-year-old child in both size and structure, shown in figure 2.7, and is 110 cm tall and weighs 20 kg.

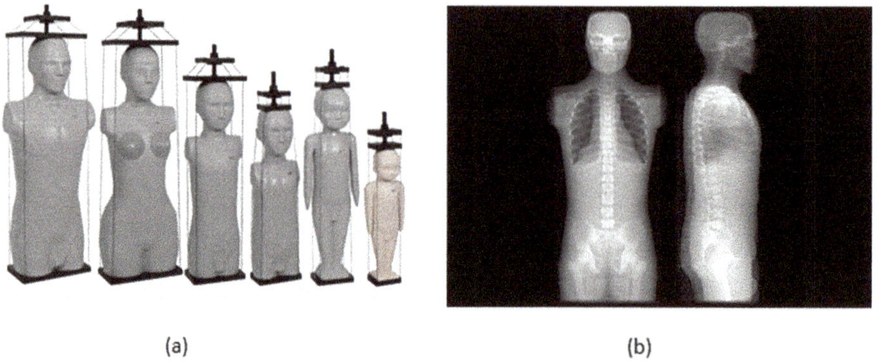

Figure 2.6. CIRS ATOM® anthropomorphic phantoms (models 701–706) www.cirsinc.com/products/all/33/atom-dosimetry-verification-phantoms/: (a) phantoms and (b) planar x-ray images of the phantoms.

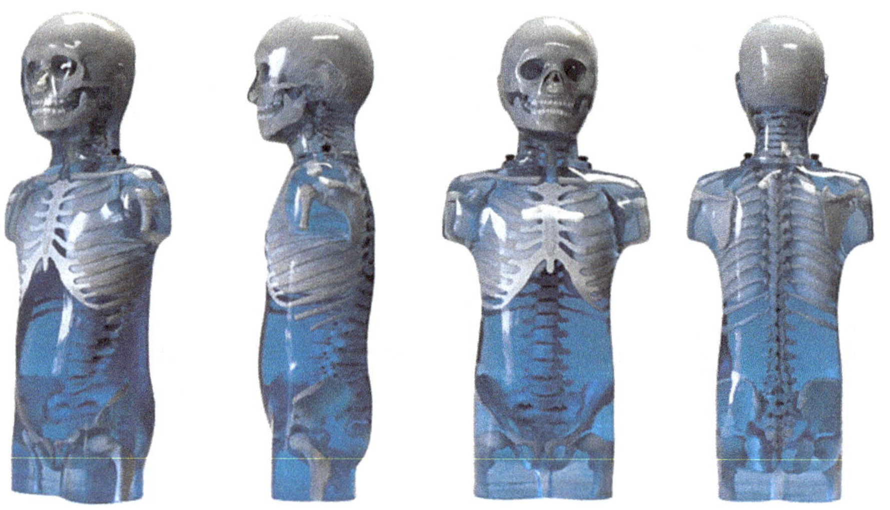

Figure 2.7. Pediatric anthropomorphic training phantom CIRS Model 715 series (http://www.cirsinc.com/products/all/36/pediatric-anthropomorphic-training-phantoms/).

The series consists of six sectional phantoms available separately or as a complete set. The complete set includes the head, trunk, right arm flexed 90° with oblique hand position, left arm extended with natural hand position, right leg flexed at 90° with natural foot position, and left leg extended with plantar flexion foot position. The most common paediatric fracture types are embedded in the left upper and lower extremities, including a buckle fracture to the tibia and a common fracture to the fibula, a fracture to the first metatarsal, a radius fracture with open reduction and hardware, and a common fracture to the second intermediate phalanx. Tissues are modelled from propietary urethane and epoxy materials that mimic the x-ray attenuation properties of human tissues for both diagnostic and therapy energy ranges from 50 keV to 25 MeV. The materials are durable, impact resistant and suitable for continuous handling. The transparent soft tissue facilitates the visual instruction of anatomical landmarks.

The phantoms are dedicated to use in training in imaging departments, as well in assisting in the monitoring and improvement of parameters and protocols common to most paediatric imaging procedures in film radiography, computed radiography, digital radiography, and CT.

During the last few years, the application of 3D printing in biomedical research has grown enormously. A basic requirement is a digital (software) model. Such models may be obtained from either 3D imaging techniques such as MRI, CT, PET, etc, or dedicated computer-aided detection (CAD) software applications. An overview of these phantoms with the used 3D printing technology is summarised in table 3.6, chapter 3.

There is a large variety of physical phantoms dedicated to dosimetry and imaging purposes which represent *part of the human body*. The following subsection focuses on anthropomorphic physical phantoms of the head, breast, and thorax.

2.3.2 Physical phantoms that model parts of the human body

2.3.2.1 Head and neck phantoms

Head and neck phantoms model the neck and the head of the human body.

The CIRS Dental and Diagnostic Head Phantom was designed for the selection, monitoring, training, and verification of scanning parameters common to most radiological procedures requiring fine anatomical details (figure 2.8). The phantom models the average male human head in both size and structure. It includes detailed 3D anthropomorphic anatomy such as brain, bone, larynx, trachea, sinus, nasal cavities, and teeth. The bones contain both cortical and trabecular separations. The teeth include a distinct dentine, enamel, and root structure including the nerve. The phantom is ideal for determining optimum system settings, commissioning new equipment, monitoring system performance, and training in dental and panoramic x-rays, as well as CT and cone-beam CT examinations. The tissue-equivalent

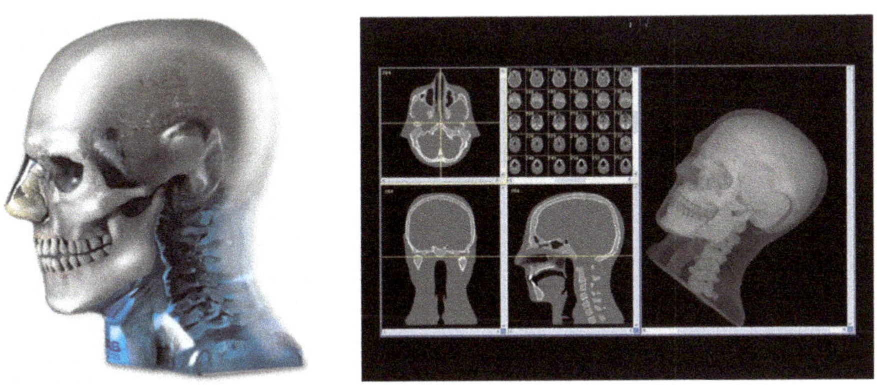

(a) (b)

Figure 2.8. The CIRS ATOM® Max Dental and Diagnostic Head Phantom is a standard of reference for diagnostic radiology of the head: (a) the physical model and (b) diagnostic x-ray images of the phantom (www.cirsinc.com/products/all/35/atom-max-dental-and-diagnostic-head-phantom/).

materials are resins mimicking the x-ray attenuation properties of human tissue for both CT and radiotherapy energy ranges between 50 keV and 25 MeV.

2.3.2.2 Torso phantoms

The *Duke phantom* (Model 07-646, Nuclear Associates, USA) is a QC phantom for conventional and digital x-ray chest systems. It provides quantitative measurements for evaluation of image quality and performance of the complete integrated digital system. The 'Duke' phantom, an x-ray image of which is shown in figure 2.9(a), was developed at the Duke University Medical Centre (Floyd and Chotas 1997). The phantom is constructed from sheets of copper and aluminium shapes resembling frontal radiographic projections of human thoracic structures (lungs, heart, ribs, and abdomen). The components are oriented and arranged to simulate an x-ray projection of a complete thorax and are then sandwiched between additional sheets of aluminium and acrylic. The top two corners of the phantom form 'shoulders', which provides a region for direct x-ray exposure of the image receptor. The 'Duke' phantom is equipped with regional test objects, a contrast-detail test pattern, a line-pair test pattern, and stainless steel wire mesh overlaying the entire chest area.

The *Alderson lung/chest phantom* (Radiology Support Device Inc., USA) is a phantom that models the human body from the neck to below the diaphragm (figure 2.9(b)). It is moulded around a male skeleton, corresponding to the external body size of a patient 175 cm tall and weighing 73.5 kg. The phantom is available either with the diaphragm permanently sealed to the interior of the phantom, fixing the lungs in place, or with a removable diaphragm, which permits the interchange of lungs and provides an unlimited number of configurations and disease simulations. The materials used to construct the phantom are equivalent to human bone and soft tissues. Animal lungs are selected to match the size of an adult male. Lungs are fixed in the inflated state and are moulded to conform to the pleural cavities of the phantom. The pulmonary arteries are injected with a blood-equivalent plastic. The phantom with a simulated left coronary artery reveals several areas of coronary

(a) (b)

Figure 2.9. Torso and lung physical phantoms: (a) x-ray image of the 'Duke' digital phantom. It is used for quality control of conventional and digital x-rays (http://www.supertechx-ray.com/Anthropomorphic/X-RayRadiography/Duke.php). (b) Lung/chest with removable diaphragm.

artery irregularity and narrowing. The different pathological conditions of the lungs are simulated by a set of custom pathologies which include different sizes of nodules and pneumonia: (a) two contiguous 0.5 cm nodules at the tip of the first left rib; (b) a 0.6 cm nodule superimposed with large vessel in the left lower lobe; (c) a 1.5 cm nodule, left mediastinal shadow; (d) a 0.6 cm nodule blending in with the right pulmonary artery; and (e) pneumonia in the right lower lobe.

2.3.2.3 Breast phantoms

Physical phantoms for x-ray breast imaging research are available in both simple geometrical forms (made of a homogeneous background with embedded test objects) and complex anthropomorphic forms.

The *Gammex 156 mammographic accreditation phantom* simulates the x-ray attenuation of a 4.2 cm slab of a compressed human breast composed of 50% adipose tissue and 50% glandular tissue (figure 2.10(a)). There are sixteen test objects in the phantom made from nylon fibrils and simulated microcalcification specs, which are of known size, shape, and density, summarised in figure 2.10(b) and described in figure 2.10(c). These represent different structures or malignancies found when imaging the breast. The phantom provides the test step needed to measure density differences.

Gammex 156 is used to identify artefacts due to grids, filters, film-processing (for film-screen systems), and digital mammography systems, as well as to detect 16 different sizes of targets, down to 0.11 mm in size, and tumour-like masses and microcalcifications.

CIRS mammographic phantoms models 010 and 011A are tissue-equivalent, anthropomorphic phantoms designed to test the performance of any mammographic system (figure 2.11). The models are realistically shaped, having compressed thicknesses ranging from 4 cm to 6 cm. In addition, they include a removable 0.5 cm adipose-equivalent tissue layer. The models consist of various glandular compositions: 30/70, 50/50, or 20/80 glandular and adipose. In addition, CIRS mammography phantoms can manufacture any glandular/adipose combination from pure gland to pure fat. Breast detail components closely simulate the radiographic properties and

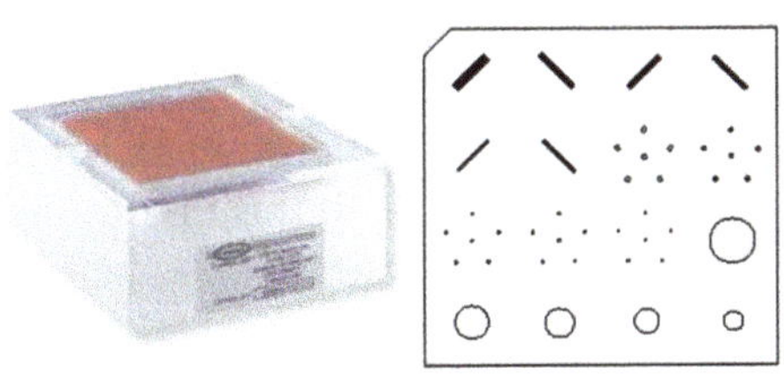

Phantom Body Material	Acrylic
Phantom dimensions	4.5x10.2x10.8 cm (HWD) 1.75x4x4.25 in)
Acrylic base	3.3cm (1.3in) thick
Cover	0.3cm (0.12in) thick
Acrylic disk	4mm thick x 1cm diameter
Test objects	Nylon fibrils (1.56, 1.12, 0.89, 0.75, 0.54 and 0.40 mm nylon fibres)
Simulated microcalcifications	(0.54, 0.40, 0.32, 0.24 and 0.16 mm specks)
Tumour-like masses	(2.00, l.00, 0.75, 0.50, 0.25)

(a) (b) (c)

Figure 2.10. GAMEX 156 breast phantom, which simulates the x-ray attenuation of a 4.2 cm slab of compressed human breast, which is composed of 50% adipose tissue and 50% glandular tissue: (a) phantom, (b) a slice with test objects, which are composed of nylon fibrils and simulated microcalcification specs, and (c) phantom specification.

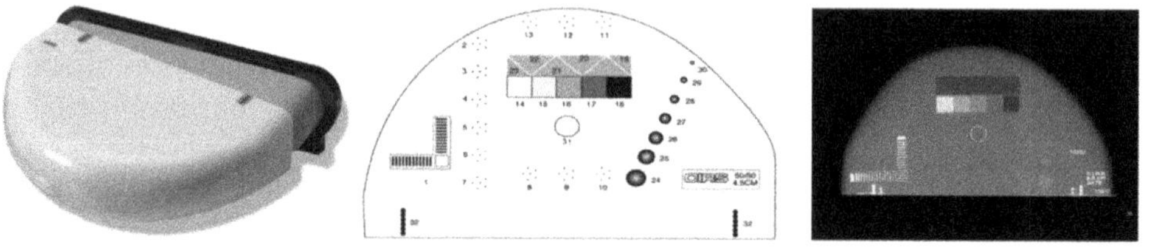
(a) (b) (c)

Figure 2.11. Tissue-equivalent phantom for mammography (model 010 and 011A): (a, b) a picture of the phantom and details, and (c) mammography image of the phantom (http://www.cirsinc.com/products/all/6/tissue-equivalent-phantom-for-mammography/).

the shapes of normal and pathological breast structures: calcifications, fibrous ducts, and tumour masses (figure 2.11(b)). Test objects range in size to allow system checks at varying levels of difficulty.

Phantoms can be used to evaluate radiation dose and image quality. Figure 2.11(c) shows an x-ray image of the physical phantom. The subjective assessment of detail visibility is easy to explore in routine clinical assessments, while densitometry analysis provides the necessary accuracy for laboratory work. The phantoms may be used for screen–film mammography or digital mammography.

Anthropomorphic phantoms were introduced in mammography to provide a more realistic task for the observer, since they mimic the composition and geometric structures of the breast, providing x-ray images similar to those present in a real breast. The first anthropomorphic breast phantom introduced for image quality and dose assessment in mammography was constructed by embedding fixed tissues in plastic (Egan and Fenn 1968). However, these types of phantoms change their characteristics with time, and they cannot be easily reproduced in identical copies.

Mammo II phantom model RS-750, shown in figure 2.12, is a patient substitute used to teach mammography positioning for craniocaudal and mediolateral oblique views; spot-compression procedures; sensitivity training for patient comfort; and is helpful in training male students (no more loaded vests to simulate breasts). The breast is mounted on a post with adjustments for height and the angles needed for various views. A small indicator box is placed next to the breast to signal when the 5 cm breast thickness is reached. An 'ouch' is represented by a red warning light which informs the trainee that a patient's pain level can be reached at about this compression. This procedure emphasises the care needed for patients when the pain zone is reached. Mammograms taken with the Mammo II phantom at a breast thickness of 5 cm present realistic images with high contrast. With this phantom, the trainees are allowed to take as many exposures as are needed to develop expertise. A moulded gel provides realistic compressibility.

Multi-modality breast biopsy and sonographic trainer CIRS 073 is designed to train users in various aspects of breast imaging and image-guided interventional procedures (figure 2.13). The phantom accurately mimics the heterogeneous appearance of breast tissue under ultrasound, mammography, and MRI. In addition, cystic and dense lesions are embedded within the breast background. Half of the dense lesions

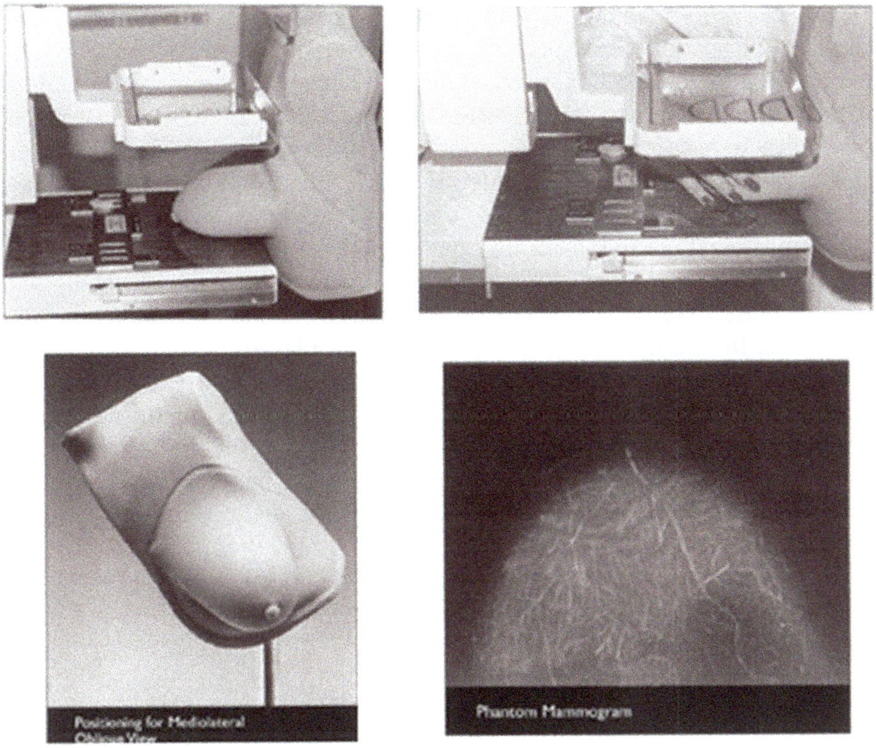

Figure 2.12. Mammo II Phantom Model RS-750 and mammography images of the phantom. (http://www.supertechx-ray.com/BreastImagingandMammography/Training/RSDRS-750.html).

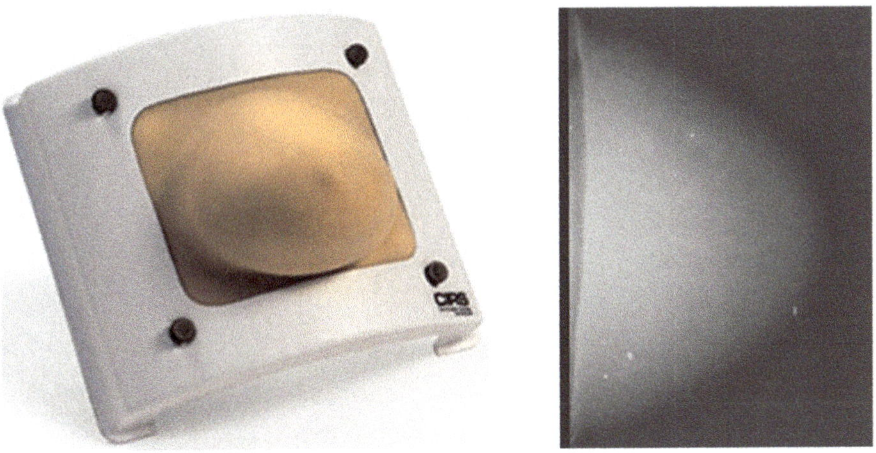

(a) (b)

Figure 2.13. Multi-modality breast biopsy and sonographic CIRS 073: (a) the phantom and (b) its x-ray projection image (http://www.cirsinc.com/products/modality/105/multi-modality-breast-biopsy-and-sonographic-trainer/).

(microcalcification) are spherical with sizes between 100 μm and 300 μm, while the other half have a spiculated shape.

The phantom includes a flexible Z-Skin™ membrane that simulates the look and feel of skin during scanning and biopsy. The skin material closes up on itself after puncture with a needle, providing good protection from desiccation even after multiple training sessions. The material inside the phantom is formulated to minimise the effect of needle tracks while practicing biopsy techniques on the embedded masses. This material also has remarkable self-healing properties, and tracks will usually disappear within minutes (sometime seconds) of needle removal. Each cystic mass may be aspirated once and each dense mass may be biopsied multiple times. In addition, the calcifications in the phantom are useful markers for image registration between modalities: x-ray, ultrasound, and MRI.

A new concept of anthropomorphic breast phantom was introduced by Caldwell and Yaffe (1990) and Yaffe *et al* (1993), who have developed the 'Rachel' phantom. This phantom consists of a breast tissue-equivalent base, simulating the tissue structures (figure 2.14(a)), combined with a mercury-enhanced mammogram, which simulates the fine details (figure 2.14(b)).

The Rachel phantom was commercially available under the name Gammex RMI, Model 169, (Gammex Inc., Middleton, WI), and was widely used for mammography image quality (IQ) and dose studies (Liu *et al* 2011), since it provides realistic breast images simulating the breast architecture and anatomical noise. The application of the Rachel phantom is limited to evaluation of 2D x-ray projection imaging systems, since this phantom does not mimic real 3D breast anatomy. It is no longer produced.

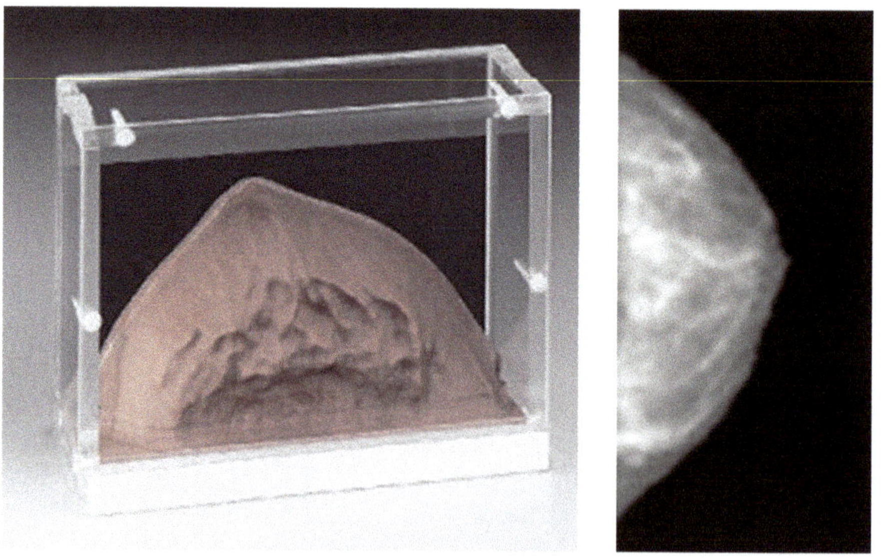

(a) (b)

Figure 2.14. 'Rachel' anthropomorphic mammo/breast phantom: (a) the phantom and (b) its projection image.

The CIRS BR3D mammography phantom is constructed with D-shaped slabs composed of a heterogeneous mixture of two breast-equivalent materials, allowing the creation of a large number of different backgrounds. The phantom, shown in figure 2.15, consists of a set of six slabs made of heterogeneous breast-equivalent material that exhibits characteristics of real breast tissue and demonstrates how underlying targets can be obscured by varying glandularity. Each slab contains two tissue-equivalent materials mimicking 100% adipose and gland tissues together in an approximate 50/50 ratio by weight. One of the slabs contains an assortment of microcalcifications, fibrils, and masses. Each semicircular-shaped slab has dimensions of 100 mm × 180 mm × 10 mm. Each slab, with its unique swirl pattern, provides varying backgrounds when arranged in different combinations and thicknesses.

This phantom is used in image quality and research studies involving breast tomosynthesis and breast CT techniques. It has the major advantage of providing a more complex background, which provides a greater challenge for target detection. Another advantage is the slab configurations, which in different combinations may provide a range of thicknesses with or without test objects.

Several anthropomorphic phantoms are constructed by combining a homogeneous background with spheres of different dimensions made of tissue-equivalent material (Baldelli *et al* 2006, Baldelli *et al* 2010, Park *et al* 2010, Taibi *et al* 2003, Cockmartin *et al* 2012). Such phantoms are depicted in figure 2.16. The phantom developed at the LUCMFR group (Catholic University of Leuven) consists of an acrylic container filled with acrylic spheres and water (figure 2.16(a)). It has been evaluated on a Siemens mammographic system with a tomosynthesis option, showing results for the

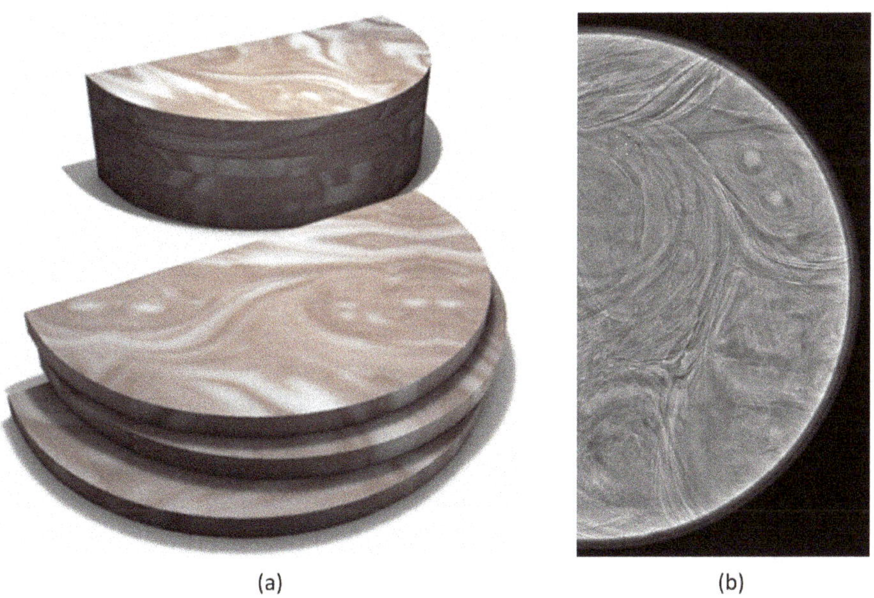

(a) (b)

Figure 2.15. The CIRS Model 020 BR3D: (a) the mammography phantom, designed to assess detectability of various size lesions within a tissue-equivalent, complex, and heterogeneous background, and (b) a mammography image.

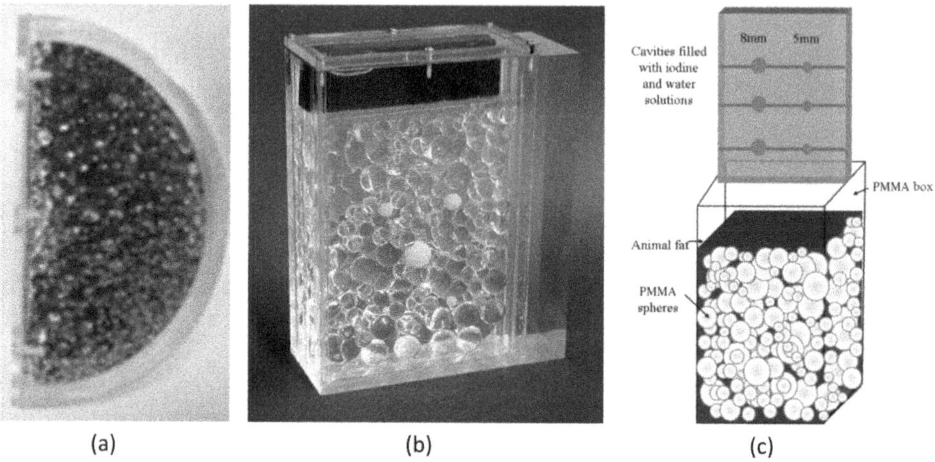

Figure 2.16. Anthropomorphic phantoms constructed by combining tissue-equivalent material spheres: (a) LUCMFR group phantom dedicated to use in 2D and 3D breast imaging, (b) photograph of the three-component phantom used for the application of the dual-energy, and (c) scheme of the three-component phantom developed for dual-energy applications. (Panels (b, c) are reproduced from Taibi *et al* 2003 and Baldelli *et al* 2006 with the permission of the publisher.)

power law exponent in the range of the exponents measured in patient data. Although the image pattern achieved with these phantoms is not similar to a real breast image, the sphere phantom is very simple to construct and can easily produce different background realisations (Freed *et al* 2011). Sphere phantoms have been widely used for evaluating the IQ and doses in mammography, and also for optimising the examination procedures of new mammographic techniques, such as dual-energy mammography, contrast-enhanced dual-energy mammography, and digital breast tomosynthesis (Argo *et al* 2004, Park *et al* 2010, Taibi *et al* 2003).

The phantom shown in figure 2.16(b) is a three-component breast phantom specifically designed and built to perform dual-energy analysis. A Plexiglas box 5 cm thick is filled with water and Plexiglas spheres of three different diameters, 8 mm, 12.7 mm, and 22.2 mm. Five polyethylene spheres are embedded in the phantom as details, with diameters ranging from 3.2 mm to 12.7 mm. A photograph of the phantom is shown in figure 2.16(c). The bigger polyethylene details are clearly visible as white spheres. The Plexiglas slab on the right of the phantom (shown in figure 2.16(b)) was used for the calculation of the incident beam intensity.

Freed *et al* (2011) developed a phantom, fabricated by mixing fresh egg white and melted refined lard, and the obtained mixture was placed in a dedicated jar with a breast-like shape. This phantom was dedicated for use in both x-ray and MRI breast imaging modalities and proved to be a useful tool for quantitative assessment of image quality in 2D and 3D mammographic techniques, and for detection and characterisation purposes (figure 2.17). This method is limited to modelling large structures in the phantom that appear considerably larger than those in the patient dataset. In addition, there is no established methodology for the production of phantoms with a desired glandularity.

Mathematical breast models are also a data source that is used to manufacture 3D anthropomorphic physical phantoms for image quality assessment of 2D and 3D breast x-ray imaging systems. Carton *et al* (2011) have developed such a physical phantom, based on a mathematical approach. Figure 2.18(a) shows printed slices which represent the glandular part of the breast, while figure 2.18(b) depicts the complete slices with realised adipose compartments. The initial evaluation of mammography and tomosynthesis images of this physical phantom (figure 2.18(c))

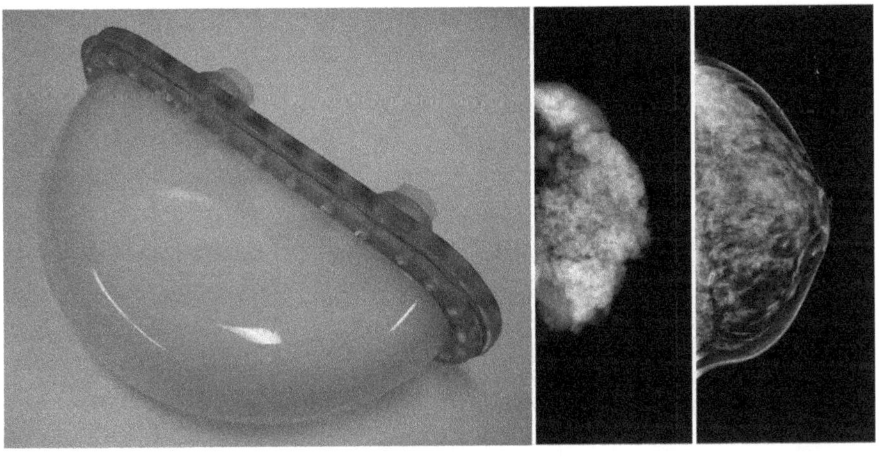

Figure 2.17. A photograph of the compressed phantom (left), example of x-ray image of phantom (middle), and example of patient mammogram (right). (Reproduced from Freed *et al* 2011 with the permission of the publisher.)

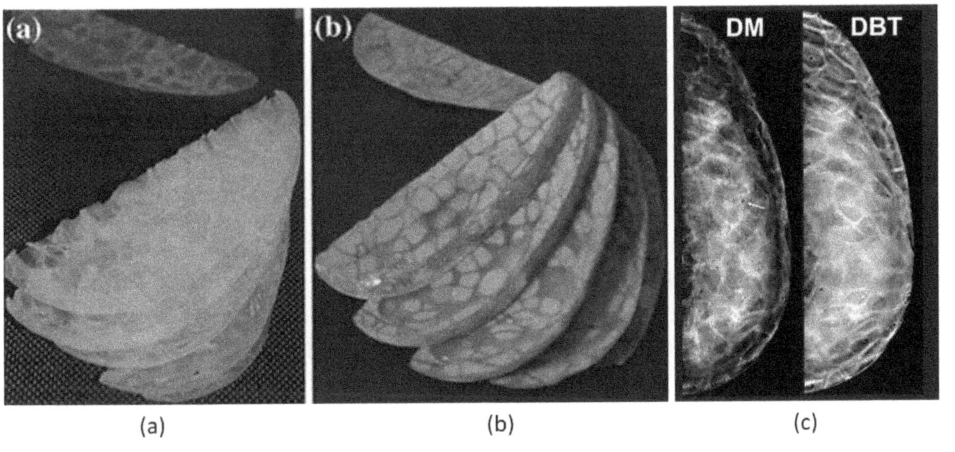

Figure 2.18. Anthropomorphic phantom developed by Carton *et al* (a) Phantom sections of the fibroglandular tissue, skin, and Cooper's ligaments composed of tissue-equivalent material with 50% glandular equivalence; (b) phantom sections after filling the structures with epoxy resin with 100% adipose equivalence; (c) digital mammography (DM) image and reconstructed digital breast tomosynthesis (DBT) image of the prototype phantom shown in (b). (Reproduced from Carton *et al* 2011 with the permission of John Wiley & Sons. © 2011 American Association of Physicists in Medicine.)

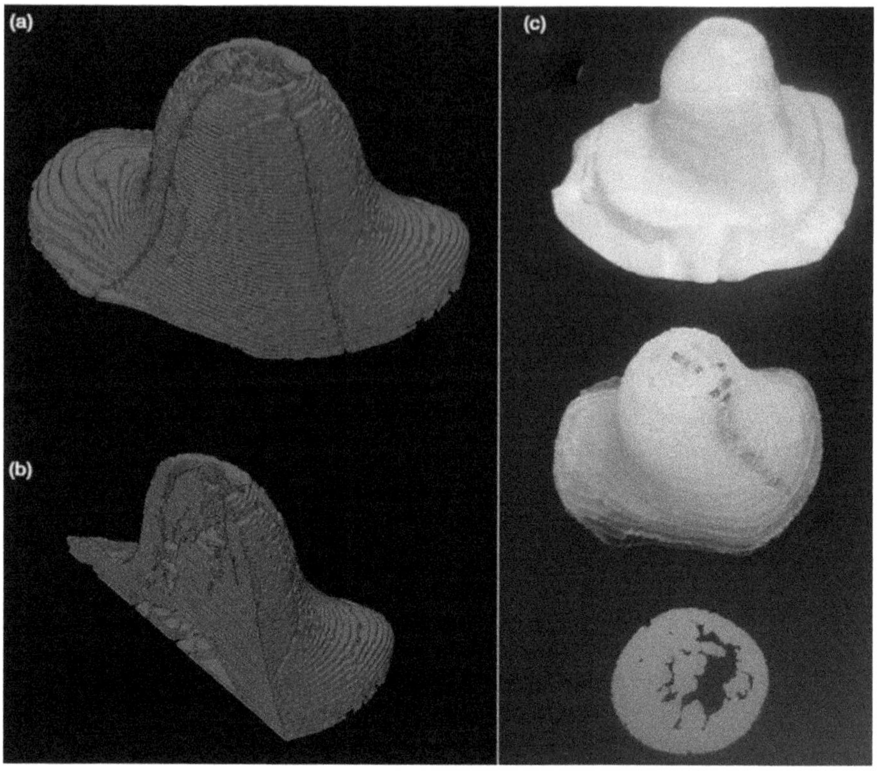

Figure 2.19. (a) CAD model of the entire phantom and (b) a cross-section through the phantom along with (c) a physical image showing the thermoplastic 'skin', stacked polyethylene breast segments, and a single breast segment of the two-compartment anthropomorphic phantom. (Reproduced from Prionas *et al* 2012 with the permission of John Wiley & Sons. © 2016 American Association of Physicists in Medicine.)

reveals its potential to be used for both qualitative and quantitative performance assessments for 2D and 3D breast x-ray imaging systems.

Further, figure 2.19 shows a two-compartment breast model, developed by Prionas *et al* (2012). The glandular part comprises water, while the adipose part is composed of polyethylene. 3D images are initially acquired at a dedicated breast CT, followed by noise removal and tissue segmentation into adipose and glandular.

Physical anthropomorphic breast phantoms may be developed by using inexpensive materials as well. Figure 2.20 shows a phantom produced from parching paper (Ikejimba *et al* 2017). The phantom was first modelled using analytical expressions and then discretized into voxels of a specified size. The interior of the breast was divided into glandular and adipose tissue classes using Voronoi segmentation, and additional structures such as blood vessels, chest muscles, and ligaments were added. The physical phantom was then fabricated from the virtual model in a slice-by-slice fashion through inkjet printing, using parchment paper and a radiopaque ink containing 33% (I33%) or 25% (I25%) iohexol by volume. This phantom is dedicated for use in full field digital mammography and digital breast tomosynthesis.

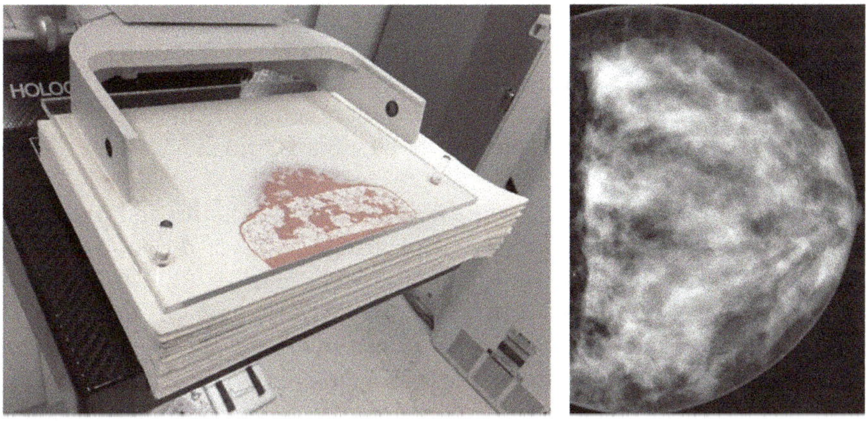

(a) (b)

Figure 2.20. Breast physical phantom built from parchment paper: (a) paper phantom and (b) its mammography image. (Reprinted from Ikejimba *et al* 2017 with the permission of John Wiley & Sons. © 2016 American Association of Physicists in Medicine.)

2.4 Anthropomorphic computational phantoms for x-ray imaging

Computational phantoms are computer models of the human body used in computerised analysis. Computer simulations of x-ray imaging techniques have become increasingly popular as computer technology has become more efficient. One of the basic components of the complete simulations in the field of x-ray imaging is the *computational phantom*. This can be as complex or as simple as the physical sample that they are intended to represent. These phantoms can simulate anything from a simple slab phantom to anatomically accurate humans.

The description of computational phantoms is based on several computational approaches: (a) geometrical primitives, (b) voxel volumes based on actual medical images of the human body, and (c) models based on non-uniform rational B-spline (NURBS) and polygon meshes. These techniques are covered in chapter 3.

The computational phantoms, similarly to the physical ones, may also be classified by their use in x-ray imaging applications, such as those dedicated to planar and 3D breast imaging, 2D radiography, lung tomosynthesis and radiography, brain and head CT, abdominal imaging, dental CBCT, etc. Phantoms also can be classified by the parts of the human body they represent: whole-body phantoms, breast phantoms, brain and head phantoms, as well as thorax phantoms, all examined in detail in the following sub-sections.

2.4.1 Whole-body computational phantoms

A number of *whole-body computational phantoms* have been developed over the years. Phantoms have been developed for a wide variety of humans, from children to adolescents to adults, male and female, as well as pregnant women. With such a variety of phantoms, many kinds of simulations can be run, from the dose received from medical imaging procedures to x-ray optimisation imaging studies and developing new techniques.

2.4.1.1 Stylised (analytical) phantoms

The stylised anthropomorphic phantoms are phantoms in which the size and the form of the human body and organs are described by mathematical expressions. Their development is considered to be a major achievement of the 1960s and accelerated the calculation of doses into body organs and tissues for 2D and 3D x-ray imaging.

The *MIRD (medical internal radiation dose) phantom* is the first anthropomorphic heterogeneous model developed at Oak Ridge National Laboratory (ORNL) dedicated to medical internal radiation dose (Snyder *et al* 1978). This model is also known as the stylised mathematical phantom, MIRD, and ORNL. The MIRD model, shown in figure 2.21(a), is composed of three basic tissues: lung, bone, and soft tissue, which are described by the corresponding material compositions listed in table 2.1. Soft tissue is used to describe the elemental composition for all the internal organs except the lung.

To represent a human body, its parts are defined using elliptical 3D primitives. For instance, the body is defined with elliptical cylinders (i.e. head, torso, and arms), while the legs are represented as truncated elliptical cones. The MIRD

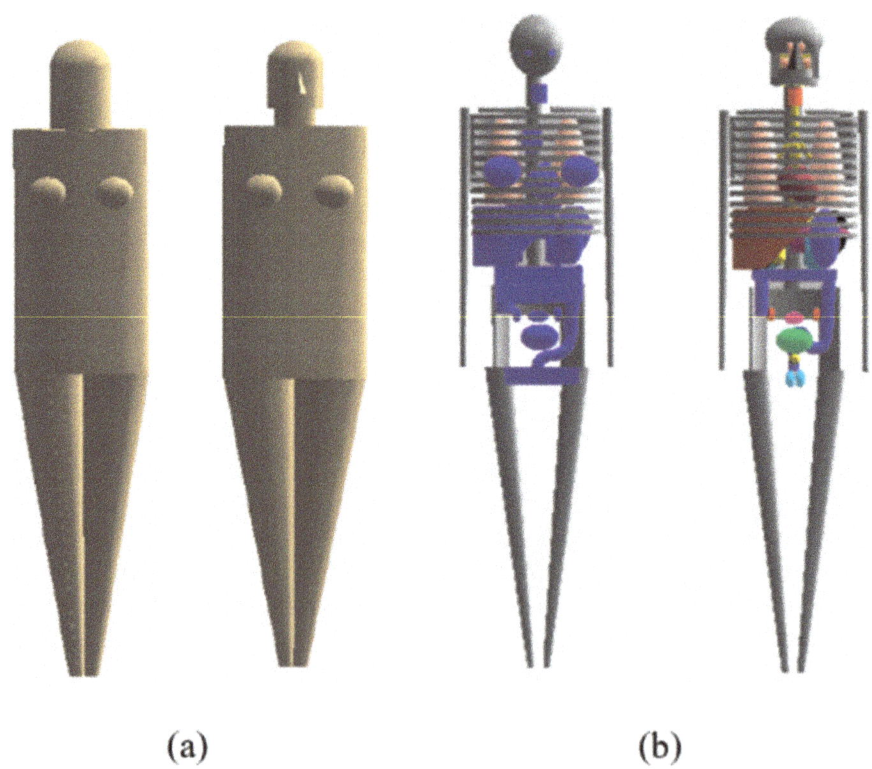

Figure 2.21. Mathematical whole-body MIRD and ORNL phantoms: (a) the exterior front shape of the MIRD-5 (left) and ORNL-UF (right) phantom models, and (b) the interior shape of the MIRD-5 (left) and ORNL-UF (right) phantom models. (Reproduced from Akkurt and Eckerman 2007. Courtesy of Oak Ridge National Laboratory, Oak Ridge, TN.)

computational models are hermaphrodite, 176 cm tall, and 73 kg in weight, corresponding to the reference values of *ICRP Publication* 89 (ICRP 2002). The phantom has the dimensions of the male reference man including testes, ovaries, and uterus, but no female breasts.

Further improvement of the MIRD phantom resulted in the development of ORNL-UF (UF—University of Florida) phantoms (Akkurt and Eckerman 2007). The ORNL-UF model, published in 1974, is composed of 23 compositions, corresponding to each internal organ (table 2.2). For both models, the arms are attached to the torso and the leg shapes are the same as in the MIRD phantom,

Table 2.1. Elemental composition and densities for the MIRD-5 phantom model.

Organ	Density, g cm^{-3}	H	C	N	O	Ca	Na	P	S	Cl	K
Lung	0.296	10.1	10.2	2.9	75.8	—	0.2	0.1	0.2	0.3	0.2
Bone	1.4	7.3	25.5	3.1	47.9	10.2	0.3	5.1	0.2	0.1	0.1
Soft tissue	1.04	10.5	22.6	2.5	63.5	—	0.1	0.1	0.2	0.1	0.2

Table 2.2. Elemental composition and densities of the major organs for the ORNL-UF phantom model.

Organ	Density, g cm^{-3}	H	C	N	O	Ca	Na	P	S	Cl	K
Lung	0.26	10.3	10.5	3.1	74.9	—	0.2	0.2	0.3	0.3	0.2
Bone	1.4	7.3	25.5	3.1	47.9	10.2	0.3	5.1	0.2	0.1	0.1
Soft tissue	1.03	10.5	25.6	2.7	60.2	—	0.1	0.2	0.3	0.2	0.2
Muscle	1.05	10.2	14.3	3.4	71.0	—	0.1	0.2	0.3	0.1	0.4
Skin	1.09	10.0	20.4	4.2	64.5	—	0.2	0.1	0.2	0.2	0.1
Brain	1.04	10.7	14.5	2.2	71.2	—	0.2	0.4	0.2	0.2	0.3
Eyes	1.07	9.6	19.5	5.7	64.6	5.1	0.1	0.1	0.3	0.1	—
Thyroid	1.05	10.4	11.9	2.4	74.5	—	0.2	0.1	0.1	0.2	0.1
Upper face	1.22	8.9	24.1	2.8	55.7	—	0.2	2.6	0.2	0.1	0.2
Larynx	1.10	9.6	9.9	2.2	74.4	—	0.5	2.2	0.9	0.3	—
Trachea	1.03	10.5	25.6	2.7	60.2	—	0.1	0.2	0.3	0.2	0.2
GI track	1.03	10.6	11.5	2.2	75.1	—	0.1	0.1	0.1	0.2	0.1
Testes	1.04	10.6	9.9	2.0	76.6	—	0.2	0.1	0.2	0.2	0.2
Ovaries	1.05	10.5	9.3	2.4	76.8	—	0.2	0.2	0.2	0.2	0.2
Uterus	1.02	10.6	31.5	2.4	54.7	—	0.1	0.2	0.2	0.2	0.2
Urinary bladder	1.04	10.5	9.6	2.6	76.1	—	0.2	0.2	0.2	0.1	0.3
Spleen	1.06	10.3	11.3	3.2	74.1	—	0.1	0.3	0.2	0.3	0.3
Heart	1.05	10.4	13.9	2.9	71.8	—	0.1	0.2	0.2	0.2	0.3
Pancreas	1.04	10.6	16.9	2.2	69.4	—	0.2	0.3	0.1	0.2	0.2
Liver	1.05	10.3	18.6	2.8	67.1	—	0.2	0.2	0.3	0.2	0.3
Kidneys	1.05	10.3	13.2	3.0	72.4	0.1	0.2	0.2	0.2	0.2	0.2
Breast	0.94	11.6	51.9	—	36.5	—	—	—	—	—	—

while the head and airway are much improved. In addition to the number of materials, the differences in the head and airway are evident. For the MIRD phantom model, the head is modelled as a simple elliptical cylinder, while in the ORNL-UF model the head is divided into parts: ellipsoids and cylinders, the eye models are revised, and sinuses and oesophagus are added.

Based on these stylised phantoms, a large number of anthropomorphic phantoms were developed: with both male and female gender specific organs for both genders representing infants and children of various ages (Cristy 1980), paediatric phantoms (Deus and Poston 1977, Hwang *et al* 1976, Jones *et al* 1976), gender specific adult phantoms ADAM and EVA (Kramer *et al* 1982), as well as a pregnant woman (Stabin *et al* 1995). Stylised models facilitate rapid dose calculation, but these phantoms are limited in their ability to represent in great realism some anatomical details. For instance, for modelling the lungs, 3D objects such as ellipsoids may be used, the design of which is far from lung realism. On the other hand, with mathematical expressions, very fine anatomic structures can be modelled.

2.4.1.2 Voxel phantoms

The development of whole-body tomographic or voxel phantoms started in the beginning of the year 2000. Compared to mathematical phantoms, voxel phantoms are true-to-nature representations of human bodies, because they are based on patient 3D images such as CT or MR images. Two of the most famous series phantoms are the MAX and FAX phantoms, developed by two research groups from the Federal University of Pernambuco, Brazil and the GSF voxel phantoms developed at the GSF Institute in Munich, Germany.

The MAX (Male Adult voXel) (Kramer *et al* 2003) and FAX (Female Adult voXel) (Kramer *et al* 2004) phantoms are based on CT images of male and female patients (figure 2.22). The MAX phantom (Kramer *et al* 2003) is based on the Zubal head–torso phantom, built from images of a male patient (Zubal *et al* 1994) as well as based on the improved MANTISSUE3-6 and VOXTISS8 phantom versions. For the construction of the MAX phantom, nine tissue compositions and densities of materials were modelled based on ICRU Report no 44 (ICRU 1989): soft tissue, adipose, lung, muscle, skin, cartilage, bone, red bone marrow, and yellow bone marrow. The height of the MAX phantom is 175.3 cm, while the body weight is 74.6 kg. The voxel size is 3.6 mm in each direction.

The FAX phantom used two set of females patient CT images. The main set of data was 151 consecutive CT images of a 37 year old female patient, with a height of 165 cm and weight of 63.4 kg. The images were recorded from scanning the trunk, the neck, and the lower part of the head including the mandible with the lower teeth. The pixel size was 0.073 cm × 0.073 cm, while the distance between two consecutive images was 0.5 cm. For the legs and the feet, an additional 206 CT images of the legs and feet of a 62 year old woman with pixel size 0.07 cm × 0.07 cm were used. For these images the distance between two consecutive images was 0.25 cm.

The original 357 images for the FAX were segmented with Paint, by manually painting every organ and tissue of interest with a different colour, which corresponded to a specific organ. Prior to this procedure, the images were processed by

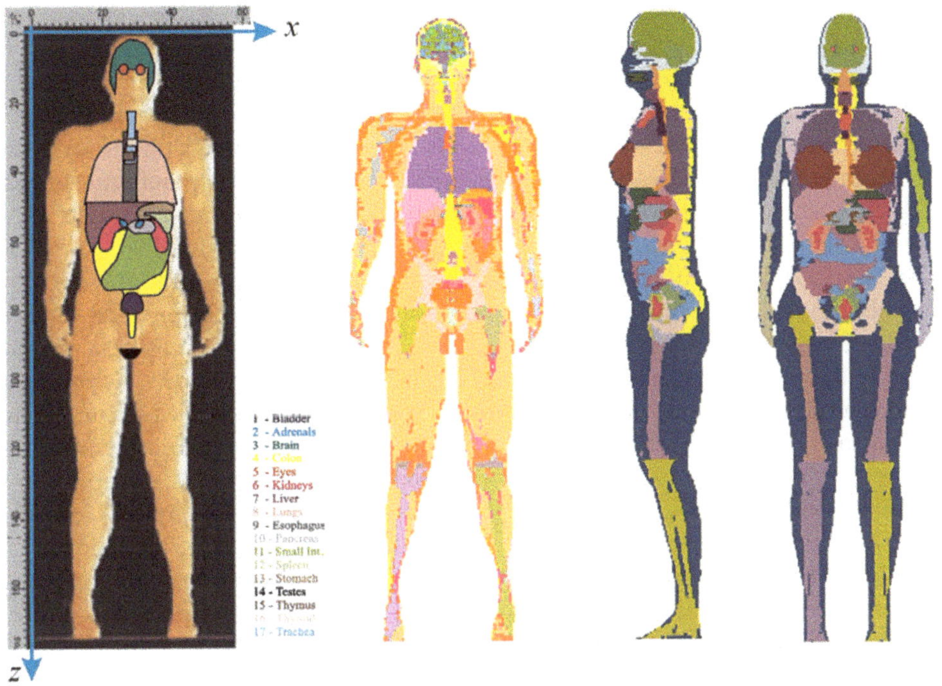

Figure 2.22. Tomographic phantoms: the MAX phantom (a) frontal view and (b) centre plane cut; the FAX phantom (c) lateral view and (d) frontal view, with skeleton and internal organs. (Reproduced from Kramer *et al* 2003, 2004 with the permission of the publisher.)

different filters in order to improve the visualisation of boundaries between organs. The following organs and tissues were segmented in the final tomographic phantom: adrenals, bladder wall, skeleton, brain, breasts, colon, kidneys, liver, lungs, muscle, oesophagus, ovaries, pancreas, small intestine, skin, spleen, stomach, thymus, thyroid, trachea, uterus, heart, and adipose tissue. Organ and tissue masses of the FAX phantom were adjusted to correspond to the anatomical specifications defined by ICRP for the female reference adult (ICRP 2002). The segmented images were resampled to achieve the same voxel size of 0.36 cm × 0.36 cm × 0.36 cm as the MAX phantom.

Further work resulted in newly segmented organs such as the gall bladder, connective tissue, salivary glands, and lymphatic nodes and the appearance of the improved MAX06 and FAX06 phantom versions.

The *GSF voxel phantoms* comprise a number of patient specific phantoms with up to 135 different organs or tissues (Petoussi-Henss *et al* 2002). The voxel matrices were obtained via segmentation from CT images from healthy volunteers and patients. The family includes 12 voxel phantoms. Amongst them are adult female, pediatric, and pregnant woman phantoms: BABY, CHILD, DONNA, FRANK, HELGA, IRENE, GOLEM, GODWIN, VISIBLE HUMAN, LAURA, KLARA, and KATJA (Petoussi-Henss *et al* 2002, Zanki *et al* 2002, Williams *et al* 1986, Fill *et al* 2004, Becker *et al* 2007). Of these, the LAURA and GOLEM phantoms were revised and as a result two new phantoms were created: REX and REGINA, which were released to the public as the ICRP adult reference male and reference female (ICRP 2009). The tomographic phantoms from the GSF family, created in 2002, were at that time the most precise representation of human anatomy to be used for computational radiation transport simulation. They replaced the mathematical (stylised) phantoms which described the human body and its internal organs using simple geometrical shapes.

Other voxel phantoms have been developed from MRI images, such as the NORMAN, NORMAN-5, and NAOMI phantoms (Dimbylow 1996, Dimbylow 2005a, 2005b). In 2001, Saito *et al* (2001) reported on the first tomographic Asian voxel phantom—the Japanese voxel male adult phantom. The segmented phantom consists of more than 100 regions enabling the calculation of doses for various parts of the body. The Japanese phantom was used with Monte Carlo techniques to calculate organ doses for external exposure due to photons and electrons up to 1 TeV (Saito *et al* 2001). Caon *et al* in 1999 constructed a tomographic phantom based on fifty-four consecutive CT scans of a 14-year-old female torso suitable for the determination of organ doses from CT (Caon *et al* 1999). The model, known as ADELAIDE's, has dimensions close to the Australian averages for her age and was used with Monte Carlo radiation transport code, specifically to calculate organ doses in head and chest CT. Smans *et al* (2008) constructed two voxel phantoms (1910 g and 590 g), representing prematurely born babies from CT and MRI. The two voxel phantoms were used in Monte Carlo calculations to assess organ doses. Further, Pan *et al* (2014) developed a 1-year-old computational phantom. The approach in developing this phantom differed from the others. The body contour was obtained from the CT images of a 1-year-old physical phantom, while the internal organs were deformed from an existing Chinese reference adult phantom. The organ locations in the 1-year-old computational phantom were manually adjusted one by one, and the organ masses were adjusted to the corresponding Chinese reference values.

The development of whole-body voxel-based phantoms accelerated the introduction of new dosimetric concepts for the calculation of equivalent dose to organs and tissues such as skin and red bone marrow (Kramer *et al* 2003). The use of these models may also result in reduction of the errors caused by scattered radiation and over-scanning effects on the dose estimation

In addition to the benefit of anatomical realism, the voxel-based phantoms are easy to use and are well connected due to the geometry of the voxel. The limitations of the voxel phantoms are related to the finite voxel resolution, which results in difficulties in modelling very thin structures such as skin, oral mucosa, the bone marrow cavities, the walls of the colon, etc, and thus the impossibility to precisely perform effective dose calculation. The voxelisation due to the use of voxels may also result in inaccurate dose calculations and low quality of simulated images. In addition, the possibility of obtaining scaled phantoms from the original voxel phantoms turns out to be limited and may lead to distortions in the representation of some organs, because of the processing of the original data on neighbouring tissues in 2D images. Scaling of such phantoms can only be performed by adjusting the voxel dimensions. Voxels, however, are homogeneous. Heterogeneity may also be possible to achieve with these models, however, this is most frequently accomplished by an algorithm during the execution of radiation transport calculations.

2.4.1.3 B-Rep phantoms

An approach to resolving the problem with scalable models is the use of Non-Uniform Rational B-Spline (NURBS) and/or polygon mesh surfaces. This approach preserves both the anatomic realism of voxel phantoms and the mathematical flexibility of constructive solid geometry (CSG) phantoms. The usual procedure in constructing such phantoms includes four steps, outlined with examples in figure 2.23. Each of these steps results in an anthropomorphic phantom represented by either B-Reps, CSG or voxels (details are given in chapter 3).

The first step is segmentation of organs and tissues. Sources of 3D images are patient medical examination from 3D imaging techniques such as PET/CT and MRI. Segmentation of the organs and tissues of interest may be performed by using well known 3D computer graphics software applications such as *3D-Doctor* (Able Software Corp, Lexington, MA), *Blender*, *ImageJ*, *Photoshop*, *Paint*, or by developing your own software application. The result of this step is segmented regions of interest.

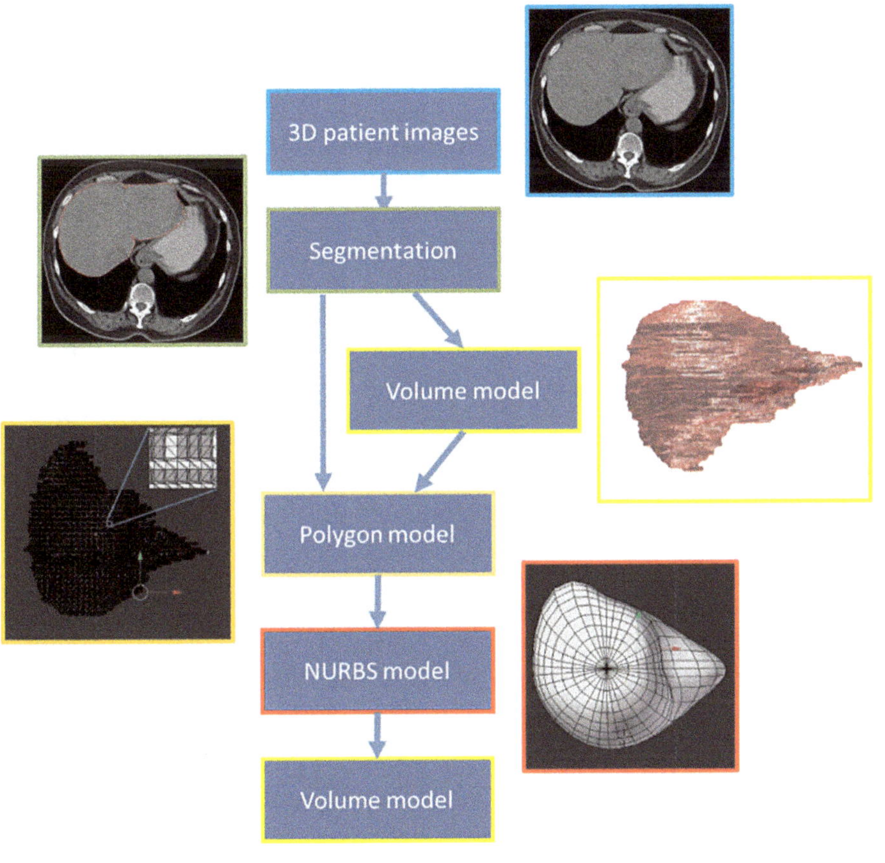

Figure 2.23. Methodology for obtaining a B-Rep phantom of the liver.

The second step is the surface reconstruction of the segmented tissues, which tissues are converted to a polygon mesh model, where the segmented organs are represented by a large array of triangular surfaces. Professional software packages (such as *Amira*, *Mimics*, *3DS Max*, *Maya*, and *Rhino* software, etc) may be used to obtain the contour of a segmented structure by applying a threshold operation. The contour of each segmented structure can then be 3D surface-rendering reconstructed via a surface generation order, and the surface of these 3D models can be smoothed via a strained or unstrained smooth order. The third step is to transform these polygon meshes into NURBS surfaces. For this step there are software tools such as *Rhinoceros* (McNeel North America, Seattle, WA), *TiniSpline* (open source C-library with bindings for various languages), etc, which may be useful for faster implementation. Finally, the NURBS-based phantom is voxelised, if needed. This step is required for compatibility of these phantoms with the radiation transport codes such as *EGS*, *MCNP*, and *GEANT*, or using in-house developed code.

A hybrid approach for building anthropomorphic phantoms is also a good alternative: to combine the polygon, NURBS, or voxelised approaches in a separate approach. Such an approach is used in the development of the hybrid anthropomorphic computational male Male Adult meSH (MASH) and Female Adult meSH (FASH) phantoms (Cassola *et al* 2010b). These phantoms are based on 3D anatomical models (which can be downloaded free on the Internet) or on organs and tissues modelled from scratch with appropriate open source software using polygon mesh surfaces. The basic software tools used in the creation of the MASH and FASH phantoms were *MakeHuman*, *Blender*, *Binvox*, and *ImageJ*. The total number of modelled organs, bones, and tissues in both phantoms is 113. Most organ and tissue masses of the voxelised versions (MAX06 and FAX06) agree with corresponding data from ICRP 89 within a margin of 2.6%. Figure 2.24(a) shows the frontal views of the FASH and MASH mesh standing adult phantoms, while figure 2.24(b) shows their voxel-based version.

These phantoms were a basis for the development of a series of computational phantoms with variable weight and height. The next-generation computational phantom models are FASH2 and MASH2, as well as MASH3 and FASH3, which are characterised by new modelled organs and tissues, such as the adipose and the glandular tissues of the breasts for the female phantom, and visceral fat separated from subcutaneous fat for the male phantom (Cassola *et al* 2010a). These are important tissues with respect to radiation risk.

Other B-Rep phantoms include the hybrid University of Florida family anthropomorphic phantoms (Lee *et al* 2007, 2008), the polygon mesh surface pair of adult male and adult female RPI-AM and RPI-AF computational phantoms (Zhang *et al* 2009), the Visible Chinese Human Adult Female (VCH-F) phantom (Liu *et al* 2005, Zhang *et al* 2008), and the hybrid 4D cardiac-torso XCAT phantom (Segars *et al* 2010, 2013). The latter presents highly detailed whole-body male and female anatomies with models for cardiac and respiratory motions (Segars *et al* 2010, 2013).

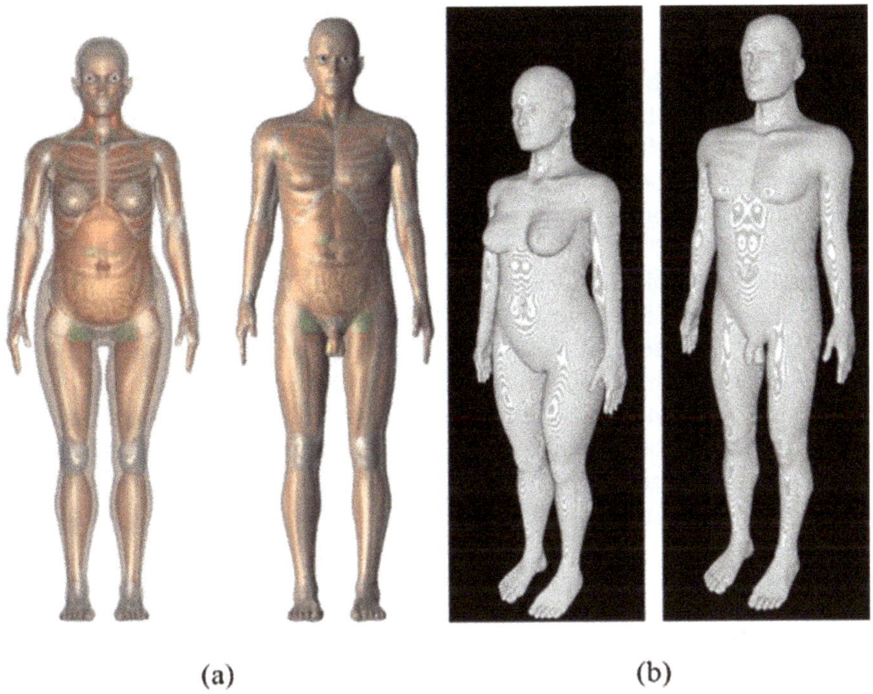

Figure 2.24. FASH and MASH whole-body phantoms: (a) frontal views of FASH and MASH, and (b) voxelised version of FASH and MASH. (Reproduced from Cassola *et al* 2010a with the permission of the publisher.)

2.4.2 Computational phantoms representing parts of the human body

2.4.2.1 Breast phantoms

The use of computational breast phantoms is required in the development and evaluation of new breast x-ray imaging modalities which aim to detect and better characterise the breast lesions in their early stage, the optimisation of new systems in clinically relevant tasks, and the performance of studies prior to implementation in clinical studies.

Simple mathematical breast phantoms are usually presented in the form of cylinders, half-ellipsoids, or slabs, and are mainly dedicated to simulation studies for dosimetry as well as optimisation of acquisition geometry. Often they are modelled from a homogeneous material (e.g. polymethyl methacrylate) that combined with the phantom thickness results in the desired glandular-to-adipose breast ratio. Such a phantom is shown in figure 2.25.

The computational breast models with realistic 3D breast tissue distribution may be classified into mathematical phantoms and phantoms created from 3D medical imaging data of patients and phantoms.

The *mathematical breast phantoms* exploit mathematical techniques to model the breast's basic elements: shape, mammary duct system, Cooper's ligaments, pectoral muscle, vessels, skin, and abnormalities. Two such examples are shown in figures

2.26 and 2.27, which correspond to the models proposed by Bliznakova *et al* (2003) and Bakic *et al* (2002, 2003). The breast elements are modelled with 3D geometrical primitives (described in chapter 3) such as ellipsoids (to represent the lobules, abnormalities), semi-ellipsoids (representing the external shape, the nipple), cylinders (to represent the ducts from the ductal system), etc. The final breast model is presented as a voxel model. Further essential improvements to these models towards anatomical and radiological realism were made by several investigators (Bliznakova *et al* 2010, Bakic *et al* 2011, Pokrajac *et al* 2012, Lau *et al* 2012). For instance, using the methodology of Pokrajac *et al* (2012) it become possible to generate phantoms with a voxel size in the range of 25–1000 μm^3/voxel. In order to overcome the voxelisation problem, several methods were proposed. One approach is to simulate smaller voxels, which, however, will increase both the phantom generation time and the memory requirements. Another approach is to account for the partial volume of various tissues in a voxel (Chen *et al* 2012). The linear x-ray attenuation coefficient of each voxel is calculated by combining attenuation coefficients proportional to the voxel sub-volumes occupied by the various tissues.

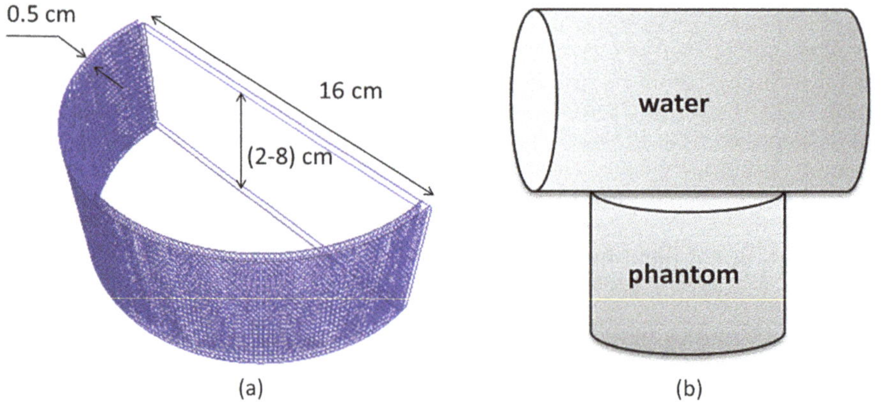

Figure 2.25. Example of a simple computational phantom used for glandular dose calculations: (a) compressed breast model used for studies in mammography and tomosynthesis, and (b) uncompressed breast model used for studies in breast CT.

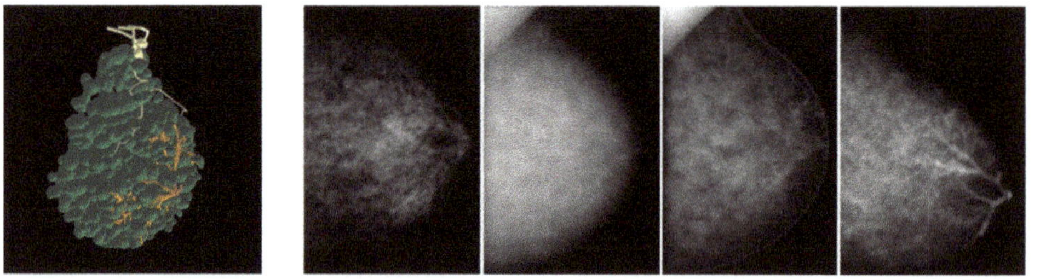

Figure 2.26. Breast model (left) and four generated projection images (right). (The projection images are reproduced from Bliznakova *et al* 2010 with the permission of John Wiley & Sons. © 2010 American Association of Physicists in Medicine.)

Based on these algorithms, several research groups generated breast models for specific needs, extending the existing models to include improved and new models of anatomical structures (Youn *et al* 2011, Chen *et al* 2011) and abnormalities such as irregular tumours and spiculated masses (Ma *et al* 2009), as well as more functionalities, such as compatibility for multimodal image simulations (Mahr *et al* 2012).

Three-dimensional geometrical primitives are the main components of the anthropomorphic breast phantom shown in figure 2.28. The computational breast is based on the physical breast developed at Catholic University of Leuven (figure 2.16(a)), which is capable of producing a structured background in both planar mammography and breast tomosynthesis. This physical breast phantom is based on the work of Gang *et al* (2010) who showed that equal volumes of differently sized acrylic spheres provide a fractal dimension of 3 as well as a power law exponent β equal to 3. The computational phantom is composed of two main parts: a semi-

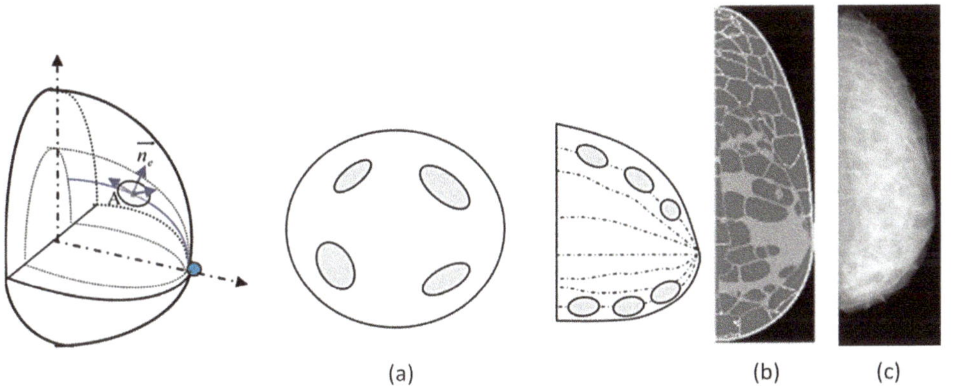

Figure 2.27. Methodology for the creation of breast phantoms. Illustration of the compartment orientation scheme at the initialisation of the region-growing procedure. (a) The compartment seeded at point A is oriented so that its shortest axis corresponds to the surface normal of an ellipsoid passing through the nipple and the seed point A. The initial distribution of the compartments is illustrated in (b) coronal and (c) vertical phantom sections, which show the central slice of a computational breast phantom and its simulated tomosynthesis projection. (Reproduced from Bakic *et al* 2011 with the permission of John Wiley & Sons.)

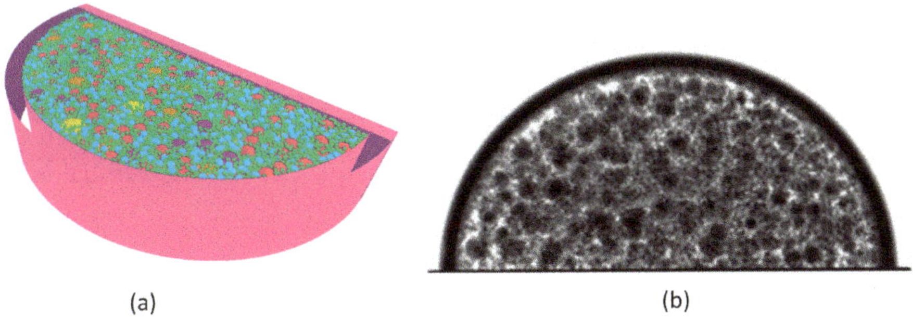

Figure 2.28. (a) Computational model of breast and (b) generated mammography image.

cylinder container and spheres of different diameters. The container is modelled as a set of four semi-cylinders and a parallelepiped. The creation of such a computational phantom is discussed in detail in chapter 4.

Anthropomorphic voxel breast models with realistic tissue distributions are also created using CT datasets, both from patients and physical phantoms. A key element in this procedure, similar to whole-body computational phantoms, is the segmentation process. This process applied to 3D medical breast imaging data results in such anthropomorphic phantoms. Using this approach, Hoeschen *et al* (2005) created voxel breast models from a high-resolution CT dataset of breast specimens, which were fixated while being compressed. These models allowed more realistic representation of the glandular tissue and enabled reasonable calculations of the average glandular dose in mammography by Monte Carlo calculations as well as simulations of different imaging procedures. A post-processed clinical dataset acquired by a breast CT system was used by Li *et al* (2009) to create a computational breast phantom. The steps followed by the researchers for generating the breast phantom are shown in figure 2.29(a), while the simulated mammography projection of the phantom is shown in figure 2.29(b). Such a design guarantees a very high degree of realism; however, it represents a single breast composition and lacks the flexibility to cover wide anatomical variations.

Other examples of voxel breast phantoms based on CT patient images and demonstrating high levels of anatomical and radiological realism, are the models developed by Huang *et al* (Huang *et al* 2011, Li *et al* 2009, Yang *et al* 2011), Segars *et al* (2014) who created one-hundred realistic and detailed 3D computational breast phantoms, and Hsu *et al* (2011, 2013). In the latter case, the computerised breast phantoms are a hybrid of two phantom generation techniques, combining empirical breast CT data with flexible computer graphics techniques. A limited set of original dedicated breast CT data was originally used to generate multiple anthropomorphic phantoms, based on image morphing techniques.

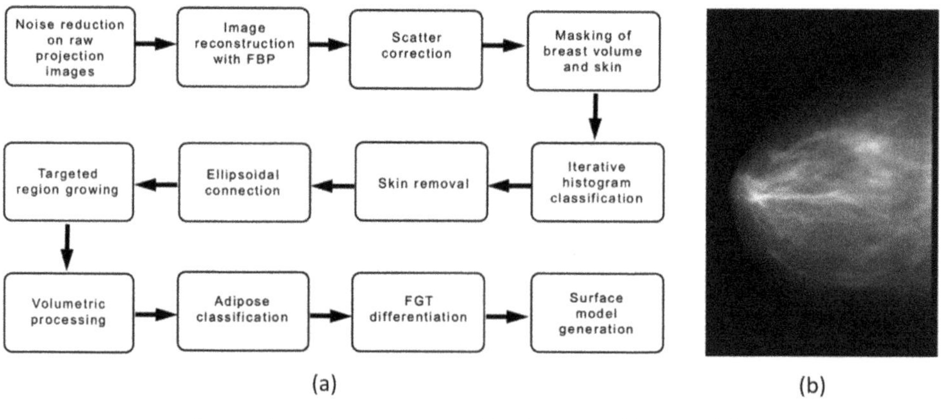

Figure 2.29. Creation of computational anthropomorphic phantoms: (a) methodology for generating a 3D computerised breast phantom from empirical data based on the work of Li *et al* (2009) and (b) a mammogram with defined attenuation coefficients that are described with three levels of fibroglandular tissue ranging from adipose to muscle. (Reproduced from Li *et al* 2009 with the permission of John Wiley & Sons.)

Computational breast models with realistic 3D breast tissue distribution and anatomical features offer a flexible, inexpensive, and fast way to investigate different aspects of 2D and 3D x-ray breast imaging: to perform accurate breast dosimetry (using Monte Carlo codes), to calculate the properties of digital mammograms, and to investigate the effect of breast tomosynthesis and CT acquisition parameters on computer-extracted texture features. These phantoms may be exploited to develop and test new CAD algorithms and to obtain realistic digital mammograms, considering the imaging characteristics of different digital mammography detectors, such as resolution and noise, and assess the breast dose for different breast imaging modalities (mammography, tomosynthesis, breast CT, and dual-energy breast imaging).

2.4.2.2 Brain and head phantoms

The 3D Shepp–Logan phantom, shown in figure 2.30, is used in various simulation studies. It consists of 12 ellipsoids with radiuses a, b, c in each direction and position defined by x, y, z and the Euler angles. These ellipsoids have different absorption properties, and resemble the outline of a head. The parameters of this phantom are given in table 2.3.

The 3D Shepp–Logan phantom continues to be used for the development of new reconstruction techniques, including general and new cone-beam CT reconstruction algorithms, as well as to validate reconstruction algorithms, such as the exact cone-beam reconstruction algorithm, the generalised filtered back-projection, and back-projection filtration methods.

High-resolution voxel human head models are also available and built from MRI and CT patient images. The Zubal CT-based head phantom (Zubal *et al* 1994) was constructed for use in Monte Carlo simulations of nuclear medicine imaging

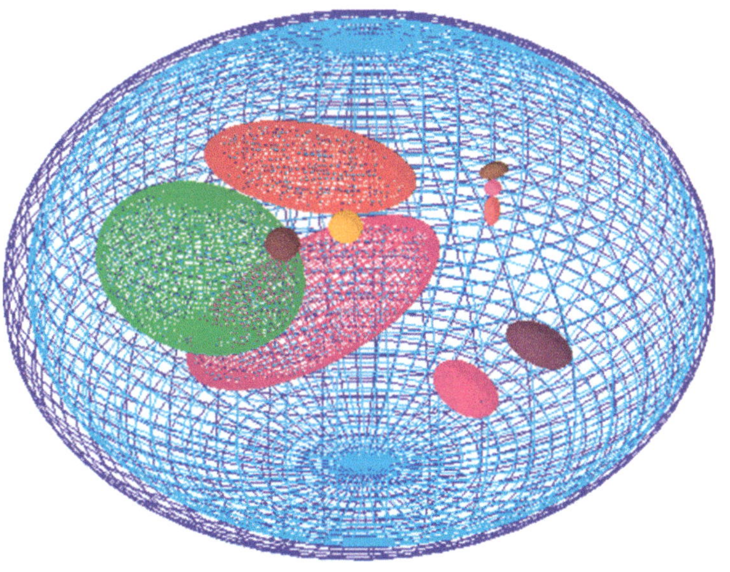

Figure 2.30. Shepp–Logan phantom, created from data listed in table 2.3.

Table 2.3. Parameters of the Shepp–Logan phantom (figure 2.30).

a	b	c	x	y	z	θ	ϕ	Density
0.69	0.9	0.92	0.0	0.0	0.0	0.0	0.0	2.0
0.6624	0.88	0.874	0.0	0.0	−0.0184	0.0	0.0	−0.98
0.41	0.21	0.16	−0.22	−0.25	0.0	0.0	72.0	−0.02
0.31	0.22	0.11	0.22	−0.25	0.0	0.0	−72.0	−0.02
0.21	0.35	0.25	0.0	−0.25	0.35	0.0	0.0	0.01
0.046	0.046	0.046	0.0	−0.25	0.1	0.0	0.0	0.01
0.046	0.02	0.023	−0.008	−0.25	−0.605	0.0	0.0	0.01
0.046	0.02	0.023	0.06	−0.25	−0.605	0.0	90.0	0.01
0.056	0.1	0.04	0.06	0.625	−0.105	0.0	0.0	0.02
0.056	0.1	0.056	0.0	0.0625	0.1	0.0	0.0	−0.02
0.046	0.046	0.046	0.0	−0.25	−0.1	0.0	0.0	0.01
0.023	0.023	0.023	0.0	−0.25	−0.605	0.0	0.0	0.01

geometries. This phantom was built by performing manual segmentation of 129 x-ray CT transverse slices of a living male human. Each voxel of the volume contains an index number designating it as belonging to a given organ or internal structure. The original x-ray CT images were of size 512 × 512 pixels with a resolution of 1 mm in the x, y plane. The z-axis resolution is 1 cm from the neck to mid-thigh and 0.5 cm from the neck to the crown of the head. This volume array represents a high-resolution model of human anatomy and can serve as a voxel anthropomorphic phantom suitable for many computer-based modelling and simulation calculations. Additionally, an MRI-based, high-resolution head phantom was also constructed (Zubal *et al* 1995). A 3D view of this phantom is shown in figure 2.31. Both phantoms and slices can be downloaded from http://noodle.med.yale.edu/zubal/data.htm.

MRI patient images were used in the creation of a realistic, high-resolution, digital, volumetric phantom of the human brain suitable for tomographic image simulation (Collins *et al* 1998). This brain phantom is made up of ten volumetric datasets that define the spatial distribution for different tissues (e.g. grey matter, white matter, muscle, skin, etc), where voxel intensity is proportional to the fraction of tissue within the voxel. Phantom construction was based on a high-resolution (1 mm isotropic voxels) low-noise dataset that was created by registering and averaging 27 T1-weighted, 12 PD-weighted, and 12 T2-weighted MRI scans from a single subject. The final volume contains 181 × 217 × 181 voxels and covers the brain completely, extending from the top of the scalp to the base of the foramen magnum. The phantom data are available from the Montreal Neurological Institute, McGill University, McConnell Brain Imaging Center website (http://www.bic.mni.mcgill.ca/brainweb).

The improved modification of this phantom has resulted in the Digital Brain Phantom II (Aubert-Broche *et al* 2006). The improvement concerns the realism of the phantom and the resulting simulations, achieved by modelling more tissue classes: the blood vessels, bone marrow, and dura mater classes. In addition, a more realistic skull class with cranium and facial bones was created (figure 2.32), which is particularly useful for SPECT, PET, and CT simulations in which bone attenuation has an important effect. In order to allow the discrimination of finer detail and enable a better classification, the voxel size of the phantom was decreased to a volume of 0.125 mm^3.

Another approach for the creation of a computational head phantom is by combining CT images of an anthropomorphic head phantom with clinically acquired MRI brain images (Riordan *et al* 2011). Such a phantom is used to assess different brain perfusion calculation methods.

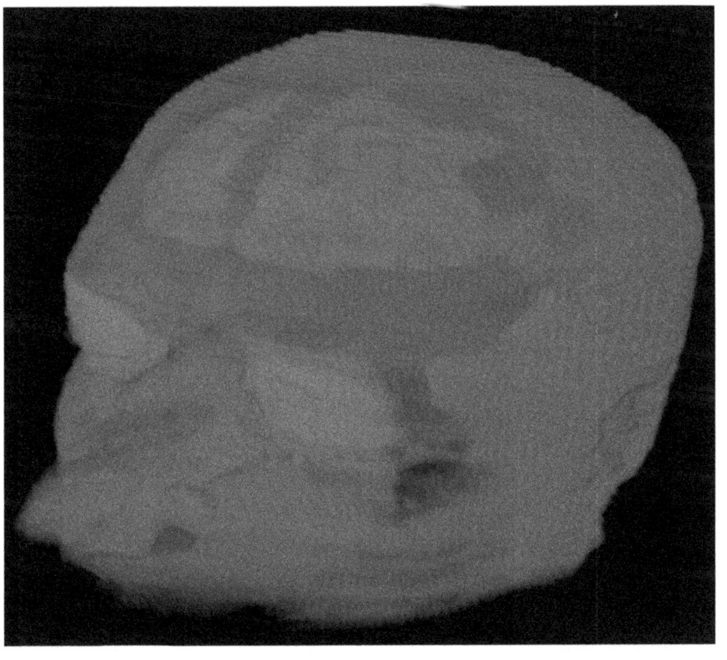

Figure 2.31. MRI-based Zubal computational head phantom.

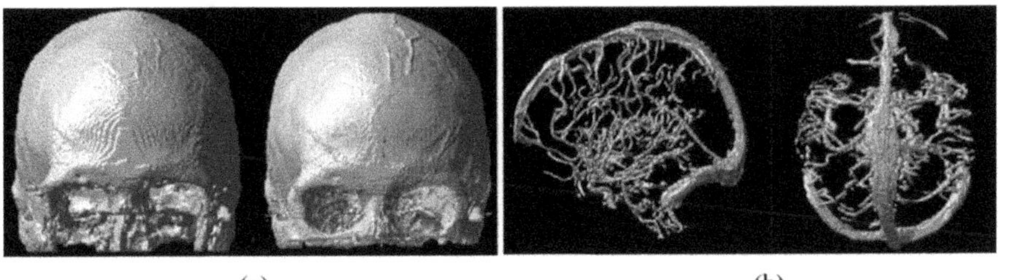

(a) (b)

Figure 2.32. A realistic computational head phantom: (a) skull classes from the previous phantom (on the left) and from the new phantom (on the right) and (b) 3D left view (on the left) and top view (on the right) of the vessels class including superficial and internal vessels. (Reproduced from Aubert-Broche *et al* 2006 with the permission of Elsevier.)

In 2014, Van Den Boom *et al* (2014) reported on a 4D digital brain phantom, created as a combination of spatial structure and tissue attenuation curves. The spatial structure of the digital phantom defines the location and types of the various anatomical structures that should be present. The anatomical structure of the digital phantom includes brain tissue separated into grey matter (GM), white matter (WM), and cerebrospinal fluid (CSF). A tissue attenuation curve was digitally added to each voxel, based on the classification of that voxel as GM, WM, or CSF. In order to divide the brain tissue into these three components (Rohlfing *et al* 2010), a 3D atlas of normal human brain anatomy was used. A Gaussian kernel ($\sigma = 1.5$ mm) is used to achieve transitions from one voxel to another voxel to look more realistic. In order to simulate an artery and a vein, two cylindrical objects were added. Finally, measured noise patterns were obtained by scanning a physical 3D skull phantom. This phantom was filled with epoxy resin that has approximately the same density as the brain tissue, to simulate noise patterns in a real human skull.

In general, voxel models of the human body are commonly used for simulating radiation dose with a Monte Carlo radiation transport code. Due to memory limitations, the voxel resolution of these computational phantoms is typically too large to accurately represent the dimensions of small features such as the eye.

The NURBS-based human brain phantom (brain portion of the extended NCAT phantom) is based on segmented MRI dataset of a normal subject, including nearly one hundred structures (http://pages.jh.edu/~rahmim/brain_phantom.html). The brain phantom was constructed using subdivision surfaces used to efficiently model structures with an arbitrary topological type, such as the brain, skull, muscle tissue, and vasculature (Rahmim *et al* 2008). The dataset consisted of 181 slices of the brain (pixel-sizes/slice-widths of 1.0 mm). 3D polygon surfaces were created by an in-house developed software application written in *Vizualisation Toolkit* (*VTK*). The *VTK* marching cubes algorithm was first used to create an initial polygon model for each structure. The polygon model was then optimised using the mesh optimisation and smoothing routines of the *VTK* software (figure 2.33).

2.4.2.3 Thoracic phantoms

Amongst the advanced thoracic phantoms is the mesh female thoracic phantom, based on the adult female reference computational phantom. Different chest girths (85, 90, 100, 110, and 120) and cup sizes (A–F) were created representing the most common thoracic female morphologies (Farah *et al* 2010). Variation of breast tissue composition and internal organ volumes with morphology were also considered. This resulted in 34 thoracic female phantoms.

Another popular phantom is the NCAT phantom, which is a NURBS-based cardiac-torso phantom including organs from visible human project CT data of the male and female. Gated MRI datasets of a normal patient and 3D angiogram data were used for motion modelling (Segars 2001).

These voxelised computational anthropomorphic phantoms are used to estimate the organ doses, effective doses, and risk indices for tested clinical chest x-ray imaging techniques (for instance chest radiography, tomosynthesis, and CT).

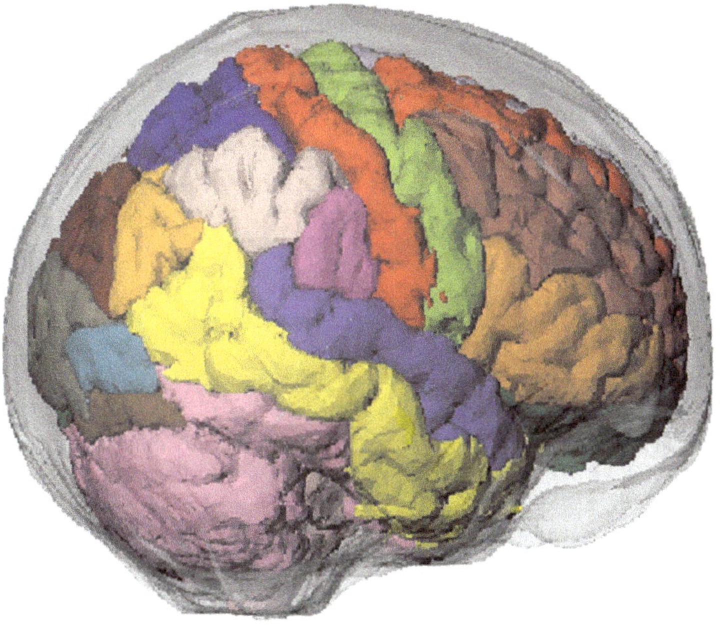

Figure 2.33. The human brain phantom with nearly 100 modelled structures in the brain (http://pages.jh.edu/~rahmim/brain_structures.html).

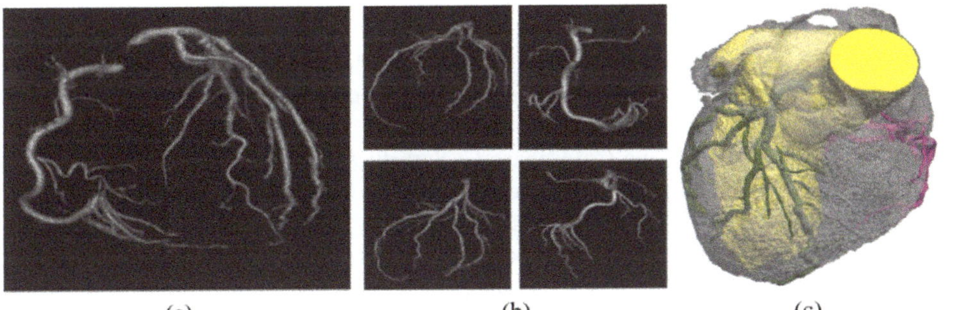

(a) (b) (c)

Figure 2.34. (a) Segmented coronary arteries; (b) views of the coronary arteries from different directions—left, two images of the right carotid artery, and right, two images of the left carotid artery; and (c) mesh-based heart phantom. (Reproduced from Gu *et al* 2011 with the permission of the publisher.)

A large family of realistic heart phantoms were reported by Gu *et al* (2011) for imaging and dosimetry purposes based on high-quality CT angiography datasets. It is based on histogram analysis of the initial image, followed by active contour segmentation of the myocardin and triangular meshes fitted to segmented high-resolution CT angiography data, as shown in figure 2.34.

2.5 Test questions

- What are the definitions of computational and physical phantoms for x-ray imaging?
- What are the special features of anthropomorphic phantoms?

- List situations/clinical tasks where anthropomorphic phantoms are preferred.
- What are the advantages of computational phantoms over physical ones?
- List situations/clinical tasks where physical phantoms are used.
- List some applications of computational anthropomorphic breast phantoms.
- List approaches for the creation of voxel phantoms.
- What are the limitations of the constructive solid geometry phantoms?

References

Abella V, Miró R, Juste B, Santos A and Verdú G 2010 Monte Carlo model of the female RANDO phantom irradiation with an Elekta Precise linear accelerator *Nucl. Instrum. Meth. Phys. Res. A* **619** 230–3

Akkurt H and Eckerman K F 2007 Development of PIMAL: mathematical phantom with moving arms and legs *ORNL report* Oak Ridge National Laboratory, Oak Ridge, TN

Argo W P, Hintenlang K and Hintenlang D E 2004 A tissue-equivalent phantom series for mammography dosimetry *J. Appl. Clin. Med. Phys.* **5** 112–9

Aubert-Broche B, Evans A C and Collins L 2006 A new improved version of the realistic digital brain phantom *NeuroImage* **32** 138–45

Bakic P R, Albert M, Brzakovic D and Maidment A D 2002 Mammogram synthesis using a 3D simulation. I. Breast tissue model and image acquisition simulation *Med. Phys.* **29** 2131–9

Bakic P R, Albert M, Brzakovic D and Maidment A D 2003 Mammogram synthesis using a three-dimensional simulation. III. Modeling and evaluation of the breast ductal network *Med. Phys.* **30** 1914–25

Bakic P R, Zhang C and Maidment A D 2011 Development and characterization of an anthropomorphic breast software phantom based upon region-growing algorithm *Med. Phys.* **38** 3165–76

Baldelli P, Bravin A, Di Maggio C, Gennaro G, Sarnelli A, Taibi A and Gambaccini M 2006 Evaluation of the minimum iodine concentration for contrast-enhanced subtraction mammography *Phys. Med. Biol.* **51** 4233–51

Baldelli P, Phelan N and Egan G 2010 Investigation of the effect of anode/filter materials on the dose and image quality of a digital mammography system based on an amorphous selenium flat panel detector *Br. J. Radiol.* **83** 290–5

Becker J, Zankl M and Petoussi-Henss N 2007 A software tool for modification of human voxel models used for application in radiation protection *Phys. Med. Biol.* **52** N195–205

Bliznakova K, Bliznakov Z, Bravou V, Kolitsi Z and Pallikarakis N 2003 A three-dimensional breast software phantom for mammography simulation *Phys. Med. Biol.* **48** 3699–719

Bliznakova K, Suryanarayanan S, Karellas A and Pallikarakis N 2010 Evaluation of an improved algorithm for producing realistic 3D breast software phantoms: application for mammography *Med. Phys.* **37** 5604–17

Busse N, Erwin W and Pan T 2013 Evaluation of a semiautomated lung mass calculation technique for internal dosimetry applications *Med. Phys.* **40** 122503

Caldwell C B and Yaffe M J 1990 Development of an anthropomorphic breast phantom *Med. Phys.* **17** 273–80

Caon M, Bibbo G and Pattison J 1999 An EGS4-ready tomographic computational model of a 14-year-old female torso for calculating organ doses from CT examinations *Phys. Med. Biol.* **44** 2213–25

Carton A K, Bakic P, Ullberg C, Derand H and Maidment A D 2011 Development of a physical 3D anthropomorphic breast phantom *Med. Phys.* **38** 891–6

Cassola V F, Kramer R, Brayner C and Khoury H J 2010a Posture-specific phantoms representing female and male adults in Monte Carlo-based simulations for radiological protection *Phys. Med. Biol.* **55** 4399–430

Cassola V F, Lima V J, Kramer R and Khoury H J 2010b FASH and MASH: female and male adult human phantoms based on polygon mesh surfaces: I. Development of the anatomy *Phys. Med. Biol.* **55** 133–62

Chen B, Shorey J, Saunders R S Jr., Richard S, Thompson J, Nolte L W and Samei E 2011 An anthropomorphic breast model for breast imaging simulation and optimization *Acad. Radiol.* **18** 536–46

Chen F Y, Pokrajac D, Shi X Q, Liu F S, Maidment A D A and Bakic P R 2012 Partial volume simulation in software breast phantoms *Proc. SPIE* **8313**

Cockmartin L, Marshall N and Bosmans H 2012a Design and evaluation of a phantom with structured background for digital mammography and breast tomosynthesis Lecture Notes in Computer Science vol 7361 (New York: Springer) pp 642–9

Collins D L, Zijdenbos A P, Kollokian V, Sied J G, Kabani N J, Holmes C J and Evans A C 1998 Design and construction of a realistic digital brain phantom *IEEE Trans. Med. Imaging* **17** 463–8

Cristy M 1980 Mathematical phantoms representing children of various ages for use in estimates of internal dose *US NRC report* NUREG/CR-1159 US Nuclear Regulatory Commission (see also Oak Ridge National Laboratory report ORNL/NUREG/TM-367)

Deus S and Poston J 1977 The development of a mathematical phantom representing a 10-yr old for use in internal dosimetry calculations *Int. J. Nucl. Med. Biol.* **4** 151

Dimbylow P 1996 The development of realistic voxel phantoms for electromagnetic field dosimetry *Workshop on Voxel Phantom Development, Chilton, UK*6–7 July 1995 pp 1–7

Dimbylow P 2005a Development of the female voxel phantom, NAOMI, and its application to calculations of induced current densities and electric fields from applied low frequency magnetic and electric fields *Phys. Med. Biol.* **50** 1047–70

Dimbylow P 2005b Resonance behaviour of whole-body averaged specific energy absorption rate (SAR) in the female voxel model, NAOMI *Phys. Med. Biol.* **50** 4053–63

Egan R L and Fenn J O 1968 Phantoms for evaluating mammography techniques and roentgenographic detail *Am. J. Roentgenol. Radium Ther. Nucl. Med.* **102** 936–40

Farah J, Broggio D and Franck D 2010 Creation and use of adjustable 3D phantoms: application for the lung monitoring of female workers *Health Phys.* **99** 649–61

Fill U A, Zankl M, Petoussi-Henss N, Siebert M and Regulla D 2004 Adult female voxel models of different stature and photon conversion coefficients for radiation protection *Health Phys.* **86** 253–72

Floyd C E Jr. and Chotas H G 1997 Anatomic phantom for evaluation of projection radiographic imaging systems *US patent* 5651046

Freed M, Badal A, Jennings R J, de las Heras H, Myers K J and Badano A 2011 X-ray properties of an anthropomorphic breast phantom for MRI and x-ray imaging *Phys. Med. Biol.* **56** 3513–33

Gang G J, Tward D J, Lee J and Siewerdsen J H 2010 Anatomical background and generalized detectability in tomosynthesis and cone-beam CT *Med. Phys.* **37** 1948–65

Gu S, Gupta R and Kyprianou I 2011 Computational high-resolution heart phantoms for medical imaging and dosimetry simulations *Phys. Med. Biol.* **56** 5845–64

Han Y *et al* 2008 Dosimetry in an IMRT phantom designed for a remote monitoring program *Med. Phys.* **35** 2519–2527

Hoeschen C, Fill U, Zankl M, Panzer W, Regulla D and Dohring W 2005 A high-resolution voxel phantom of the breast for dose calculations in mammography *Radiat. Protect. Dosim.* **114** 406–9

Hsu C M, Palmeri M L, Segars W P, Veress A I and Dobbins J T 3rd 2011 An analysis of the mechanical parameters used for finite element compression of a high-resolution 3D breast phantom *Med. Phys.* **38** 5756–70

Hsu C M, Palmeri M L, Segars W P, Veress A I and Dobbins J T 3rd 2013 Generation of a suite of 3D computer-generated breast phantoms from a limited set of human subject data *Med. Phys.* **40** 043703

Huang S Y, Boone J M, Yang K, Packard N J, McKenney S E, Prionas N D, Lindfors K K and Yaffe M J 2011 The characterization of breast anatomical metrics using dedicated breast CT *Med. Phys.* **38** 2180–91

Hwang J, Shoup R and Poston J 1976 Mathematical description of a newborn human for use in dosimetry calculations *ORNL report* ORNL/TM-5453 Oak Ridge National Laboratory, Oak Ridge, TN

ICRP 2009 Adult reference computational phantoms *Ann. ICRP* **39** 1–166

ICRP 2002 Basic anatomical and physiological data for use in radiological protection reference values, ICRP Publication 89 *Ann. ICRP* **32** 3–4

ICRU 1989 Tissue substitutes in radiation dosimetry and measurement ICRU report 44 International Commission on Radiation Units and Measurements, Bethesda, MD

Ikejimba L C, Graff C G, Rosenthal S, Badal A, Ghammraoui B, Lo J Y and Glick S J 2017 A novel physical anthropomorphic breast phantom for 2D and 3D x-ray imaging *Med. Phys.* **44** 407–16

Jones R, Poston J, Hwang J, Jones T and Warner G 1976 The development and use of a fifteen-year-old equivalent mathematical phantom for internal dose calculations *ORNL report* ORNL/TM-5278 Oak Ridge National Laboratory, Oak Ridge, TN

Kramer R, Khoury H J, Vieira J W, Loureiro E C, Lima V J, Lima F R and Hoff G 2004 All about FAX: a Female Adult voXel phantom for Monte Carlo calculation in radiation protection dosimetry *Phys. Med. Biol.* **49** 5203–16

Kramer R, Vieira J W, Khoury H J, Lima F R and Fuelle D 2003 All about MAX: a male adult voxel phantom for Monte Carlo calculations in radiation protection dosimetry *Phys. Med. Biol.* **48** 1239–62

Kramer R, Zankl M, Williams G and Drexler G 1982 The calculation of dose from external photon exposures using reference human phantoms and Monte Carlo methods Part I: the male (ADAM) and female (EVA) adult mathematical phantoms Report GSF-Bericht-S 885 Gesellschaft für Strahlenund Umweltforschung, Munich

Lau B A, Reiser I, Nishikawa R M and Bakic P R 2012 A statistically defined anthropomorphic software breast phantom *Med. Phys.* **39** 3375–85

Lee C, Lodwick D, Hasenauer D, Williams J L and Bolch W E 2007 Hybrid computational phantoms of the male and female newborn patient: NURBS-based whole-body models *Phys. Med. Biol.* **52** 3309–33

Lee C, Lodwick D, Williams J L and Bolch W E 2008 Hybrid computational phantoms of the 15-year male and female adolescent: Applications to CT organ dosimetry for patients of variable morphometry *Medi. Phys.* **35** 2366–82

Li C M, Segars W P, Tourassi G D, Boone J M and Dobbins J T 3rd 2009 Methodology for generating a 3D computerized breast phantom from empirical data *Med. Phys.* **36** 3122–31

Liu Q, Gong H and Luo Q 2005, *Proc. Ann. Int. Conf. of the IEEE Engineering in Medicine and Biology* **vol 7** (Piscataway, NJ: IEEE) pp 5172–5

Liu X, Lai C J, Whitman G J, Geiser W R, Shen Y, Yi Y and Shaw C C 2011 Effects of exposure equalization on image signal-to-noise ratios in digital mammography: a simulation study with an anthropomorphic breast phantom *Med. Phys.* **38** 6489–501

Ma A K, Gunn S and Darambara D G 2009 Introducing DeBRa: a detailed breast model for radiological studies *Phys. Med. Biol.* **54** 4533–45

Mahr D M, Bhargava R and Insana M F 2012 Three-dimensional *in silico* breast phantoms for multimodal image simulations *IEEE Trans. Med. Imaging* **31** 689–97

Pan Y, Qiu R, Gao L, Ge C, Zheng J, Xie W and Li J 2014 Development of 1-year-old computational phantom and calculation of organ doses during CT scans using Monte Carlo simulation *Phys. Med. Biol.* **59** 5243–60

Park S, Jennings R, Liu H M, Badano A and Myers K 2010 A statistical, task-based evaluation method for three-dimensional x-ray breast imaging systems using variable-background phantoms *Med. Phys.* **37** 6253–70

Petoussi-Henss N, Zankl M, Fill U and Regulla D 2002 The GSF family of voxel phantoms *Phys. Med. Biol.* **47** 89–106

Pokrajac D D, Maidment A D and Bakic P R 2012 Optimized generation of high resolution breast anthropomorphic software phantoms *Med. Phys.* **39** 2290–302

Prionas N D, Burkett G W, McKenney S E, Chen L, Stern R L and Boone J M 2012 Development of a patient-specific two-compartment anthropomorphic breast phantom *Phys. Med. Biol.* **57** 4293–307

Rahmim A *et al* 2008 Accurate event-driven motion compensation in high-resolution PET incorporating scattered and random events *IEEE Trans. Med. Imag.* **27** 1018–1033

Riordan A J, Prokop M, Viergever M A, Dankbaar J W, Smit E J and De Jong H W A M 2011 Validation of CT brain perfusion methods using a realistic dynamic head phantom *Med. Phys.* **38** 3212–21

Rohlfing T, Zahr N M, Sullivan E V and Pfefferbaum A 2010 The SRI24 multichannel atlas of normal adult human brain structure *Hum. Brain Mapp.* **31** 798–819

Saito K, Wittmann A, Koga S, Ida Y, Kamei T, Funabiki J and Zankl M 2001 Construction of a computed tomographic phantom for a Japanese male adult and dose calculation system *Radiat. Env. Biophys.* **40** 69–76

Segars W P, Tsui B M W, Lalush D, Frey E, King M A and Manocha D 2001 Development and application of the new dynamic NURBS-based cardiac-torso (NCAT) phantom *J. Nucl. Med.* **42** 23

Segars W P *et al* 2013 Population of anatomically variable 4D XCAT adult phantoms for imaging research and optimization *Med. Phys.* **40**

Segars W P, Sturgeon G, Mendonca S, Grimes J and Tsui B M 2010 4D XCAT phantom for multimodality imaging research *Med. Phys.* **37** 4902–15

Segars W P, Veress A I, Wells J R, Sturgeon G M, Kiarashi N, Lo J Y, Samei E and Dobbins J T 2014 Population of 100 realistic, patient-based computerized breast phantoms for multi-modality imaging research *Proc. SPIE* **9033** 90331X

Smans K, Tapiovaara M, Cannie M, Struelens L, Vanhavere F, Smet M and Bosmans H 2008 Calculation of organ doses in x-ray examinations of premature babies *Med. Phys.* **35** 556–68

Snyder W S, Ford M R and G. W G 1978 Estimates of specific absorbed fractions for monoenergetic photon sources uniformly distributed in various organs of a heterogeneous phantom *J. Nucl. Med.* **Suppl 3** 7–52

Stabin M, Watson E, Cristy M, Ryman J, Eckerman K, Davis J, Marshall D and Gehlen K 1995 Mathematical models and specific absorbed fractions of photon energy in the nonpregnant adult female and at the end of each trimester of pregnancy ORNL report No. ORNL/TM-12907 Oak Ridge National Laboratory, Oak Ridge, TN

Struelens L, Vanhavere F and Smans K 2008 Monte Carlo model of the female RANDO phantom irradiation with an Elekta Precise linear accelerator *Phys. Med. Biol.* **53** 5831–44

Taibi A, Fabbri S, Baldelli P, di Maggio C, Gennaro G, Marziani M, Tuffanelli A and Gambaccini M 2003 Dual-energy imaging in full-field digital mammography: a phantom study *Phys. Med. Biol.* **48** 1945–56

Van Den Boom R, Manniesing R, Oei M T H, Van Der Woude W J, Smit E J, Laue H O A, Van Ginneken B and Prokop M 2014 A 4D digital phantom for patient-specific simulation of brain CT perfusion protocols *Med. Phys.* **41**

Williams G, Zankl M, Abmayr W, Veit R and Drexler G 1986 The calculation of dose from external photon exposures using reference and realistic human phantoms and Monte Carlo methods *Phys. Med. Biol.* **31** 449–52

Yaffe M J, Byng J W, Caldwell C B and Bennett N R 1993 Anthropomorphic radiological phantoms for mammography *Med. Prog. Technol.* **19** 23–30

Yang X, Sechopoulos I and Fei B 2011 Automatic tissue classification for high-resolution breast CT images based on bilateral filtering *Proc. SPIE* **7962** 79623H

Yi Y, Lai C J, Han T, Zhong Y, Shen Y, Liu X, Ge S, You Z, Wang T and Shaw C C 2011 Radiation doses in cone-beam breast computed tomography: a Monte Carlo simulation study *Med. Phys.* **38** 589–97

Youn H, Han J C, Cho M K, Jang S Y, Kim H K, Kim J H, Tanguay J and Cunningham I A 2011 Numerical generation of digital mammograms considering imaging characteristics of an imager *Nucl. Instrum. Meth.* A **652** 810–4

Zanki M, Fill U, Petoussi-Henss N and Regulla D 2002 Organ dose conversion coefficients for external photon irradiation of male and female voxel models *Phys. Med. Biol.* **47** 2367–85

Zhang G, Luo Q, Zeng S and Liu Q 2008 The development and application of the visible Chinese human model for Monte Carlo dose calculations *Health Phys.* **94** 118–25

Zhang J, Na Y H, Caracappa P F and Xu X G 2009 RPI-AM and RPI-AF, a pair of mesh-based, size-adjustable adult male and female computational phantoms using ICRP-89 parameters and their calculations for organ doses from monoenergetic photon beams *Phys. Med. Biol.* **54** 5885–908

Zubal I G, Harrell C R, Smith E O, Rattner Z, Gindi G and Hoffer P B 1994 Computerized three-dimensional segmented human anatomy *Med. Phys.* **21** 299–302

Zubal I G *et al* 1995 Two dedicated software, voxel-based anthropomorphic (torso and head) phantoms, *Int. Conf. at the National Radiological Protection Board, Chilton, UK*

Chapter 3

Design and composition of anthropomorphic phantoms

3.1 Introduction

The previous chapter, chapter 2, introduced the *anthropomorphic phantoms* as real or virtual objects composed of tissue-equivalent materials, aiming to provide a realistic representation of the body's anatomy and properties for use in various real or virtual studies in research or applications. Realistic representation means to mimic the outer shape and/or internal structures of the modelled human tissue, and the properties mainly depend on the properties of the *tissue substitute* used, which is any material that simulates a body of tissue in its interaction with ionising radiation.

In this chapter, attention is paid to the tissue substitutes used in virtual phantoms dedicated to tasks in diagnostic radiology and interventional procedures. The aim is to summarise the basic materials used to mimic real human tissues, and specifically their elemental composition, physical density, and effective atomic number. For physical phantoms, the fabrication of such materials is outlined as well. This chapter also focuses on the phantom geometries and approaches for their modelling. At the end of the chapter test questions and quizzes are provided.

3.2 Physical anthropomorphic phantoms

The phantom materials dedicated to mimic human tissues must have physical and radiological properties that represent the physical and radiological properties of the real tissues as accurately as possible. Among the physical properties, the important ones include the physical density and the effective atomic number (Z_{eff}), which both can be used as assessments of a material's tissue equivalence. Among the important radiological properties of the phantom materials are the electron density and both the mass attenuation (μ/ρ) and mass–energy-absorption (μ_{en}/ρ) coefficients. The

doi:10.1088/2053-2563/aae197ch3

electron density provides more insight into how a material will behave in a radiation field, while the mass–energy-absorption coefficient indicates how much energy is deposited locally in the tissue (Pantelis *et al* 2004, Hill *et al* 2005). The mass attenuation coefficient reflects the imaging characteristics, i.e. how easily the x-ray beam can penetrate through the material. A mathematical description of the x-ray linear attenuation coefficient (Qi *et al* 2010) is given as:

$$\mu = \rho \frac{N_A}{A}(\sigma_{sc} + \sigma_{PE}), \tag{3.1}$$

where N_A, A, and ρ are the Avogadro's number, the atomic mass, and the density, respectively, and σ_{PE} and σ_{sc} are the photoelectric and scattering cross sections. The scattering cross section σ_{sc} is described as:

$$\sigma_{sc} = Z(F_{KN}(E) + F_{coh}(Z, E)), \tag{3.2}$$

where $F_{KN}(E)$ and $F_{coh}(Z, E)$ are the Klein–Nishina (Compton) cross section and the coherent scattering cross section, respectively. The photoelectric cross sections σ_{PE} can be approximated as:

$$\sigma_{PE} \sim \frac{Z^m}{E^3}, \tag{3.3}$$

where m is a constant. The effective atomic number Z_{eff} and Z_{eff}^{PE} are calculated following the approach of Jones *et al* (2003) as follows:

$$Z_{eff} = \sum_i w_i Z_i,\ Z_{eff}^{PE} = \sqrt[3.5]{\sum_i a_i Z_i^{3.5}}\ a_i = [(w_i Z_i / A_i) / \sum_i (w_i Z_i / A_i)], \tag{3.4}$$

where w_i, Z_i, and A_i are the mass fraction, atomic number, and mass number, respectively. Z_{eff} is defined as a mass-weighted average of the elemental atomic numbers, while Z_{eff}^{PE} is defined in Johns and Cunningham (1983) as a more representative quantity for photoelectric absorption. Jones *et al* (2003) showed that during the development of the tissue substitutes, both Z_{eff} and Z_{eff}^{PE} are good predictors of how well the tissue-equivalent materials' mass attenuation coefficients and mass–energy-absorption coefficients, respectively, would match those of the reference tissues (such as soft tissue, lung tissue, skeleton, etc).

There are a limited number of available materials which combine both physical and radiological properties similar to those of human tissues. Table 3.1 summarises the elemental composition of the most popular materials used in the fabrication of physical phantoms for x-ray imaging. It is, however, important to point out that the radiological properties of these materials depend on the used energy range, which suggests that materials used for applications in the low diagnostic energy range, for instance the mammography range, may not be suitable tissue-equivalent materials for applications at other higher diagnostic energy ranges (for instance CT). The same is valid for the x-ray modalities: materials used in radiology may not be suitable as tissue-equivalent materials in MRI or ultrasound.

This table contains data from several sources. The main data source is the publication of White *et al* (1977) who published a list of materials suitable for

Table 3.1. Elemental composition (by mass) as percentage weights, density, effective atomic number, and electronic density of materials used as tissue substitutes.

Materials	H	C	N	O	F	Mg	Si	Al	Cl	Ca	Density, g cm^{-3}	Z_{eff}	Z_{eff}^{PE}
PMMA[a]	0.08	0.60		0.32							1.19	6.24	6.56
Polyethylene	0.144	0.856									0.93	5.28	5.53
Paraffin wax	0.149	0.851									0.93	5.25	5.51
Polycarbonate	0.055	0.756		0.189							1.20	6.10	6.33
Polystyrene soft tissue equivalent[a]	0.077	0.923									1.06	5.62	5.74
Liquid water	0.120			0.880							1.0	7.16	7.47
Polyurethane	0.094	0.642	0.031	0.233							1.03–1.08	6.03	6.35
Nylon[a]	0.098	0.637	0.124	0.141							1.14	5.92	6.21
Epoxy resin	0.070	0.740		0.190							1.10	6.03	6.29
CB1-epoxy based[c]	0.089	0.722	0.056	0.132					0.001		1.11	5.85	6.19
CB2-epoxy based[c]	0.095	0.713	0.030	0.162					0.001		1.10	5.90	6.23
CB3-epoxy based[c]	0.113	0.742	0.016	0.120					0.009		1.02	5.79	6.45
CB4-epoxy based[c]	0.079	0.6838	0.030	0.206					0.002		1.15	6.07	6.43
BR10 (breast)[c]	0.092	0.72	0.05	0.128					0.001	0.011	0.95	6.02	7.07
BR12 (breast)[c]	0.087	0.699	0.024	0.179					0.001	0.01	0.97	6.10	7.08
LN4 (lung)[c]	0.086	0.69	0.023	0.18					0.001	0.025	0.60	6.34	7.97
MS10—muscle equivalent[c]	0.091	0.71	0.043	0.13					0.001	0.025	0.99	6.21	7.91
MS11—muscle equivalent[c]	0.084	0.68	0.023	0.19					0.001	0.023	0.99	6.32	7.89
MS15—muscle equivalent[c]	0.10	0.63	0.009	0.16				0.10	0.005		1.00	6.61	7.77
MS20—muscle equivalent[c]	0.08	0.58	0.018	0.19		0.13			0.001		1.00	6.78	7.74
SK1 (suitable for skin)[c]	0.074	0.60	0.045	0.11	0.12			0.055	0.001		1.10	6.68	7.45
AP1 (adipose)[c]	0.09	0.70	0.042	0.11	0.063				0.001		0.92	6.05	6.39
AP3 (adipose)[c]	0.11	0.74	0.012	0.10	0.038				0.007		0.90	5.89	6.46
AP6 (adipose)[c]	0.08	0.69	0.024	0.17	0.031				0.001		0.92	6.04	6.42
AP7 (adipose)[c]	0.084	0.691	0.023	0.169	0.031				0.0014		0.92	6.05	6.42

(Continued)

SB3 (hard cortical bone)[c]	0.031	0.31	0.01	0.38			0.001	0.27	1.84	10.42	13.92
SB4 (mimics hard cortical bone)[c]	0.037	0.32	0.01	0.36			0.003	0.27	1.75	10.36	13.91
SB5 (mimics hard cortical bone)[c]	0.026	0.306	0.010	0.389			0.001	0.27	1.87	10.46	13.94
IB1—inner bone[c]	0.079	0.638	0.0423	0.099			0.142		1.16	7.41	9.91
STES—soft-tissue-equivalent substitutes[b]	0.072	0.592	0.022	0.218	0.093	0.01	0.001		1.04	6.69	7.53
STES-NB (new born)[b]	0.07	0.581	0.021	0.223	0.094	0.01	0.001		1.04	6.77	7.65
BTES—bone-tissue-equivalent substitutes[b]	0.04	0.378	0.015	0.353		0.119	0.010	0.094	1.40	8.80	11.48
BTES-NB[b]	0.051	0.462	0.019	0.302		0.07	0.018	0.078	1.22	8.22	10.95
LTES—lung-tissue-equivalent substitutes[b]	0.07	0.574	0.021	0.224	0.093	0.017	0.001		0.30	6.83	7.77
Water equivalent (wt1)	0.084	0.680	0.023	0.189			0.001	0.024	1.00	6.33	7.94
Soft tissue—wt1-lung	0.084	0.680	0.023	0.189			0.001	0.026	1.00	6.37	8.05
LN10 (lung)—same as wt1	0.084	0.680	0.023	0.189			0.001	0.026	0.25–0.35	6.37	8.05
Bees wax[d]	0.136	0.812	0.001	0.052						5.43	5.71
Red articulation wax[d]	0.139	0.861								5.31	5.54
Orange articulation wax[d]	0.143	0.857								5.29	5.53
Paraffin I[d]	0.152	0.848								5.46	5.51
Paraffin II[d]	0.151	0.842	0.002	0.005						5.26	5.52
Bolus[d]	0.141	0.841		0.018						5.33	5.59
Nylon[d]	0.098	0.621	0.119	0.162					1.16	5.95	6.26
PMMA[d]	0.081	0.598	0.001	0.320					1.16	6.24	6.56
Pitch[d]	0.093	0.735		0.172						5.88	6.17

[a] NIST database.
[b] Jones *et al* (2003).
[c] White *et al* (1978).
[d] Ferreira *et al* (2010).

reproducing the physical and radiological properties of tissues such as adipose, breast, muscle, lung, and bone. The proposed materials marked as AP (adipose), BR (breast), MS (muscle), LN (lung), and SK (skin) are all manufactured with an epoxy resin base CB1, CB2, CB3, CB4, and CB5 (listed in table 3.1) with the addition of other materials such as polyethylene, calcium carbonate, aluminium oxide, and magnesium oxide to adjust the mass density. As an example, these phantom materials are used in the manufacture of the commercially available Leeds test objects (www.leedstestobjects.com/) as well as CIRS phantoms (for instance BR3D, figure 2.15, chapter 2).

Jones *et al* (2003) introduced five new materials based on an epoxy resin base and the technology proposed by White *et al* (1977) for use in the construction of tomographic or image-based phantoms for organ dose assessment in paediatric radiology.

The production of these materials is summarised in table 3.2, while their elemental composition is given in table 3.1: STES-NB, BTES-NB, LTES, STES, and BTES, materials which radiographically mimic the soft tissues, skeletal tissues, and lung tissues of new-born patients (with the abbreviation NB) and patients at ages of 1 year and older. One example of anthropomorphic whole-body phantoms built from urethane-based STES and LTES is shown in figure 3.1. The modified LTES and STES were developed by the group of Winslow *et al* (2009).

The last nine rows in table 3.1 show the elemental composition of bolus, nylon, orange articulation wax, red articulation wax, PMMA, bees wax, two types of paraffin (paraffin I, paraffin II), pitch, and water, which were tested as substitutes for human brain tissue in dosimetry for diagnostic radiology (Ferreira *et al* 2010). The table reveals that water is the best substitute for brain among the materials investigated, while PMMA, a material often used for the manufacture of head phantoms for CT, cannot be considered to be a suitable substitute for human brain tissue in dosimetry.

Table 3.2. Material composition of the University of Florida (UF) tissue-equivalent substitutes (Jones *et al* 2003).

	Material composition (% by mass)				
Constituent	STES-NB	BTES-NB	LTES	STES	BTES
Araldite GY6010	50.7	44.6	49.4	51.8	36.4
Jeffamine T-403	20.3	17.9	19.8	20.7	14.6
Polyethylene	8.0		8.0	8.0	
Magnesium oxide	15.5		15.5	15.5	
Silicon dioxide	2.0	15.0	2.0	0.5	25.5
Phenolic microspheres	3.5		3.5	3.5	
Calcium carbonate		19.5			23.5
Polyvinyl chloride		3.0			
DC 1107[a]			0.8		
DC 200/50[a]			1.0		

[a] DC 1107 and DC 200/50 are a foaming agent and a surfactant, respectively, used to inflate LTES.

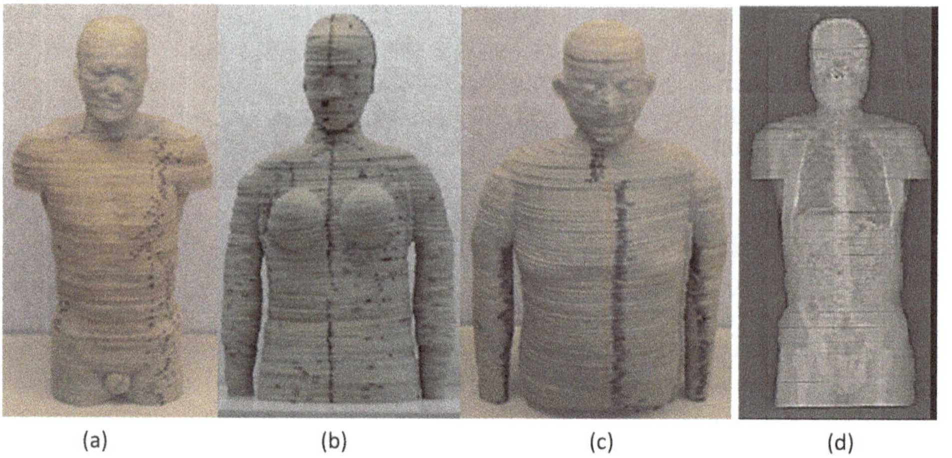

Figure 3.1. (a) Phantom based on a segmented CT dataset of an adult male; (b) phantom based on a computational adult female hybrid dataset; (c) phantom based on a computational adult male hybrid dataset; and (d) a CT topogram of a tomographic physical phantom. (Reproduced from Winslow *et al* 2009.)

Tables 3.3 and 3.4 concern liquid-tissue-surrogate (LTS) materials proposed by FitzGerald *et al* (2017) that accurately represent human soft tissues in terms of density and x-ray attenuation coefficients. The method for their production is simple and the proportions are listed in table 3.3. The evaluated mixtures of water, glycerol, butanol, methanol, sodium chloride, and potassium nitrate were intended to emulate human adipose, blood, brain, kidney, liver, muscle, pancreas, and skin (table 3.3). The hypothetical elemental composition, density, and Z_{eff} of these materials is shown in table 3.4.

The production of anthropomorphic phantoms is ultimately related to the use of new fabrication 3D printing technologies. They offer the creation of more complex and anatomically realistic breast phantoms. Currently, the existence of low-cost 3D printing devices gives an opportunity to simplifying the production and supply of realistic anthropomorphic phantoms that can be tailored to individual needs and represent different patient shapes. This process is greatly assisted by the development of software tools dedicated to medical imaging.

The elemental composition of materials available with low-cost 3D printing devices (discussed in chapter 8) is listed in table 3.5. The elemental composition (weighted by fraction) for the candidate printing materials used for skin, gland, and adipose is from the study of Ivanov *et al* (2018). These materials are used with stereolithography and fused deposition modelling (Ivanov *et al* 2018).

Table 3.6 summarises some of the developed physical phantoms with the use of 3D printing technology dedicated to x-ray research.

The design and the construction of various physical phantoms are already discussed in chapter 2.

Table 3.3. Formulation of LTS materials.

	Compound mass fraction (%)					
	Glycerine	Butanol	Methanol	NaCl	KNO_3	Water
LTS adipose #3	35.5	64.5				
LTS adipose #2	40.5	59.5				
LTS blood	26.3		2.9	0.8	1.1	68.9
LTS brain	27.9		10.7	0.9	0.8	59.8
LTS kidney	28.4		7.1	0.7	0.8	63.0
LTS liver	28.0		9.0	0.6	1.2	61.2
LTS muscle	29.0		8.0	0.4	1.3	60.0
LTS pancreas	27.9		16.7	0.5	0.6	54.3
LTS skin	47.6		8.0	0.6	0.4	43.9
LTS spleen	27.5		3.2	0.8	0.8	67.7

Table 3.4. Characteristics of the proposed LTS materials.

		Elemental mass fraction (%)							
	Density g mL^{-1}	H	C	N	O	Na	Cl	K	Z_{eff}
LTS adipose #3	0.93	11.9	55.7		32.4				6.34
LTS adipose #2	0.95	11.6	54.4		34.0				6.38
LTS blood	1.06	10.4	11.4	0.2	76.9	0.3	0.5	0.4	7.53
LTS brain	1.04	10.5	14.9	0.1	73.3	0.4	0.5	0.3	7.45
LTS kidney	1.05	10.4	13.8	0.1	74.7	0.3	0.4	0.3	7.44
LTS liver	1.06	10.4	14.3	0.2	74.0	0.2	0.4	0.5	7.46
LTS muscle	1.05	10.4	14.3	0.2	74.2	0.2	0.2	0.5	7.45
LTS pancreas	1.04	10.6	17.2	0.1	71.4	0.2	0.3	0.2	7.32
LTS skin	1.09	10.0	21.6	0.1	67.5	0.2	0.4	0.2	7.26
LTS spleen	1.06	10.4	12.0	0.1	76.4	0.3	0.5	0.3	7.48

Table 3.5. Elemental composition (weighted by fraction) for the candidate printing materials used for skin, gland, and adipose from the study of Ivanov *et al* (2018).

Material	H	C	N	O[a]	Density g cm^{-3}
PMMA_LAB	0.082±0.002	0.656±0.002		0.262±0.006	1.190
Clear	0.085±0.002	0.648±0.002	0.059±0.002	0.208±0.006	1.180
Flex	0.087±0.002	0.639±0.002	0.044±0.002	0.230±0.006	1.137
Gray	0.084±0.002	0.638±0.002	0.052±0.002	0.226±0.006	1.175
PET-G	0.052±0.002	0.685±0.002	0.012±0.002	0.251±0.006	1.236
PLA	0.058±0.002	0.541±0.002	0.018±0.002	0.383±0.006	1.250
ABS	0.078±0.002	0.862±0.002	0.059±0.002	0.001±0.006	1.019
Nylon	0.097±0.002	0.656±0.002	0.114±0.002	0.133±0.006	1.111

[a] The amount of oxygen was obtained by subtracting the sum of all other elements from 1.

3.3 Software anthropomorphic phantoms: tissue simulating materials, geometry

Computational phantoms are mathematical descriptions or matrix models of human organs or tissues of the body, organ systems, or the whole body, which aim to accurately mimic a given characteristic or a set of characteristics of the represented parts of the body. They are mainly used in virtual clinical experiments, proof of concept studies, quantitative estimation tasks, and for educational purposes. Some application examples include:

- Obtaining synthetic but realistic medical images (using different imaging modalities).
- Testing the feasibility of new imaging modalities.
- Computer calculations for estimation of the dose delivered to the entire body and to specific organs.
- Design and assessment of shielding against occupational exposure.
- Estimation of the dose to other parts of the body when a diagnostic or therapeutic dose is delivered to a specific site.
- Radiation transport calculations for computing a nuclear gamma camera image and evaluation of site-specific doses due to the radioactive agent.

When designing a computational phantom for a given purpose, the following considerations must be taken into account:

- Reproducing the anatomy at a sufficient level.
- Correctly modelling the interaction of the radiation with the matter of the specific organs/tissues.
- Achieving computational efficiency.
- Assuring geometrical compatibility with the Monte Carlo code that eventually carries out the radiation transport calculation.

To simulate the radiological properties of the tissue of interest, a detailed composition modelling of the tissue is needed. Elemental compositions and densities of various human tissues are defined by ICRU report 44 (1989) (ICRP 2002) and some of them are summarised in table 3.7.

Experimentally measured data are also used in simulations.

Geometry design

Three basic geometric approaches are used for the design of a computational anthropomorphic phantom, as outlined in figure 3.2: solid geometry, B-Reps, and voxel geometry. Hybrid approaches, which are combinations of the basic ones, are also applicable.

Constructive solid geometry (CSG). CSG operates on geometrical primitives from a certain predefined family of basics shapes which include quadrics such as cylinders, ellipsoids, cones, semi-ellipsoid, or sets of polygons such as cuboids, pyramids,

Table 3.6. Developed physical phantoms with 3D printing technology. Images are reprinted from the corresponding reference with the permission of the publisher.

Modelled human tissue	Materials and printer technology used	Phantom image (reproduced with the permission of the publisher of the corresponding manuscript)	Application	Reference
Breast	Inkjet-based Objet260 Connex3 printer with *VeroCyan* material.	 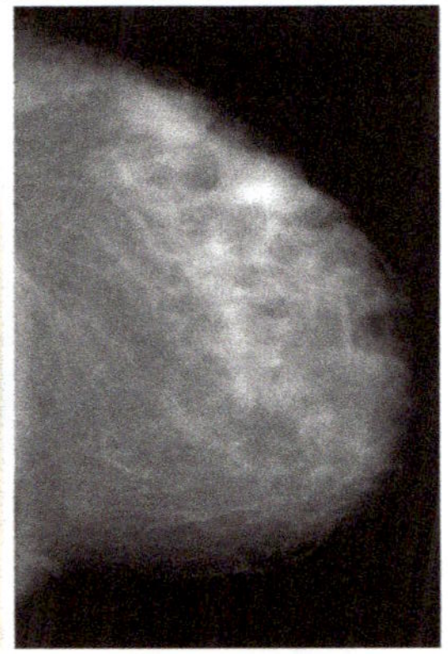	Development of collections of representative patient models that could be used to assess the effect of anatomical variability on system performance, hence making bench testing studies a step closer to clinical trials.	Badal *et al* (2018)
Thorax	Nylon (PA2200 (Polyamide 12), ρ of laser sintered part = 0.93 g cm^{-3}, bulk ρ = 0.45 g cm^{-3})—used for the airways, lung blood vessels, outer lung surface, and tumours. Printed with selective laser sintering. Gypsum (ZCorp® zp151 high-performance powder, ρ = 1.57 g cm^{-3})—used to mimic the bony structures. Printed with an Inkjet printer (layer thickness = 0.1 mm, 600 × 540 dpi, Zcorp 650®, 3D Systems).	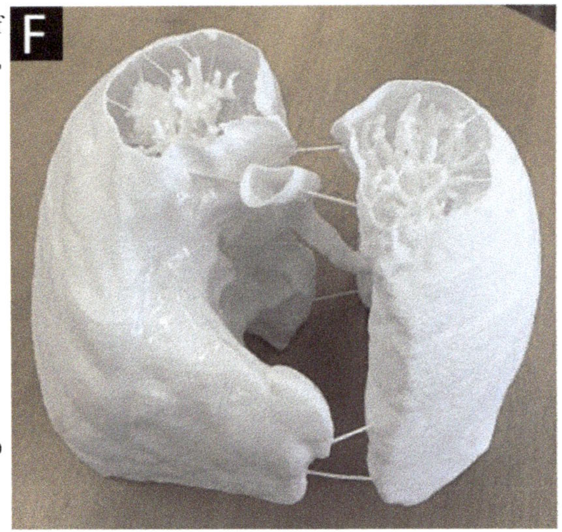	The phantom closely resembles a real patient in terms of spatial accuracy and is currently being used to evaluate x-ray-based imaging quality and positional verification techniques for radiotherapy.	Hazelaar *et al* (2018)

(*Continued*)

Full scale phantom	PLA filament with Gigabot 2.0 printer	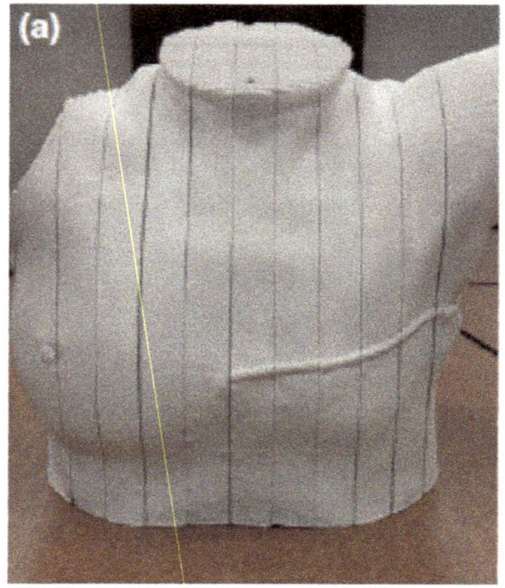	Radiotherapy and interventional procedures.	Craft and Howell (2017)
CT phantoms	Conventional inkjet technology (HP Deskjet 6940) in combination with modified, iodine-enhanced ink. Paper thickness = 0.1 mm.	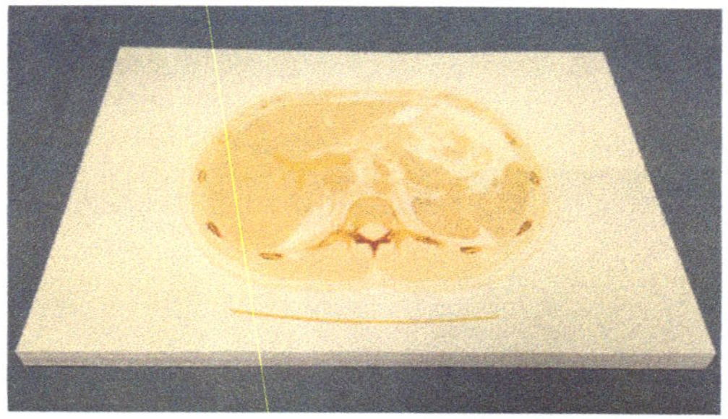	Stacking of printed images presented an opportunity to create 3D objects with desired shape and CT attenuation.	Jahnke *et al* (2017)
Breast	Inkjet printing, using parchment paper and a radiopaque ink containing 33% (I33%) or 25% (I25%) iohexol by volume.	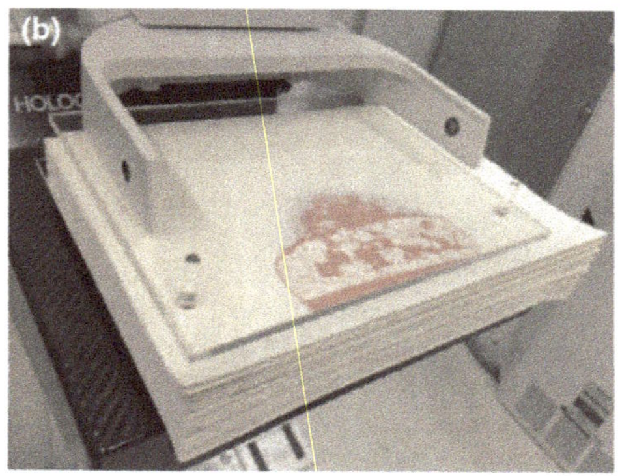	Evaluation of novel and existing 2D and 3D breast imaging techniques.	Ikejimba *et al* (2017)

Breast	Objet500 Connex printer. Materials chosen for the phantoms: TangoGray, VeroWhite, and TangoPlus.		Evaluation of novel and existing 2D and 3D breast imaging techniques.	Kiarashi *et al* (2015)
Breast	Polyjet Eden printer deposits liquid FC-720 photopolymer, layer by layer, according to the model of the fibroglandular tissue region.		Evaluation of novel and existing 2D and 3D breast imaging techniques.	Carton *et al* (2011)

Table 3.7. Elemental compositions and densities of various human tissues (ICRP 2002).

Organ	H	C	N	O	Ca	Na	P	S	Cl	K	Density, g cm^{-3}
Breast	11.6	51.9	—	36.5	—	—	—	—	—	—	0.94
Lung	10.3	10.5	3.1	74.9	—	0.2	0.2	0.3	0.3	0.2	0.26
Bone	7.3	25.5	3.1	47.9	10.2	0.3	5.1	0.2	0.1	0.1	1.4
Brain	10.7	14.5	2.2	71.2	—	0.2	0.4	0.2	0.3	0.3	1.04
Soft tissue	10.5	25.6	2.7	60.2	—	0.1	0.2	0.3	0.2	0.2	1.03
Muscle	10.2	14.3	3.4	71.0	—	0.1	0.2	0.3	0.1	0.4	1.05
Eyes	9.6	19.5	5.7	64.6	—	0.1	0.1	0.3	0.1	—	1.05
Pancreas	10.6	16.9	2.2	69.4	—	0.2	0.3	0.1	0.2	0.2	1.04
Kidneys	10.3	13.2	3.0	72.4	0.1	0.2	0.2	0.2	0.2	0.2	1.05
Liver	10.3	18.6	2.8	67.1	—	0.2	0.2	0.3	0.2	0.3	1.05
Heart	10.4	13.9	2.9	71.8	—	0.1	0.2	0.2	0.2	0.3	1.05
Spleen	10.3	11.3	3.2	74.1	—	0.1	0.3	0.2	0.2	0.3	1.06
Urinary bladder	10.5	9.6	2.6	76.1	—	0.2	0.2	0.2	0.2	0.3	1.04
Uterus	10.5	31.5	2.4	54.7	—	0.1	0.2	0.2	0.1	0.2	1.02
Ovaries	10.5	9.3	2.4	76.8	—	0.2	0.2	0.2	0.2	0.2	1.05
Testes	10.6	9.9	2.0	76.6	—	0.2	0.1	0.2	0.2	0.2	1.04
GI Tract	10.6	11.5	2.2	75.1	—	0.1	0.1	0.1	0.2	0.1	1.03
Trachea	10.5	25.6	2.7	60.2	—	0.1	0.2	0.3	0.2	0.2	1.03
Larynx	9.6	9.9	2.2	74.4	—	0.5	2.2	0.9	0.3	—	1.10
Upper face	8.9	24.1	2.8	55.7	5.1	0.2	2.6	0.2	0.1	0.2	1.22
Thyroid	10.4	11.9	2.4	74.5	—	0.2	0.1	0.1	0.2	0.1	1.05
Skin	10.0	20.4	4.2	64.5	—	0.2	0.1	0.2	0.3	0.1	1.09

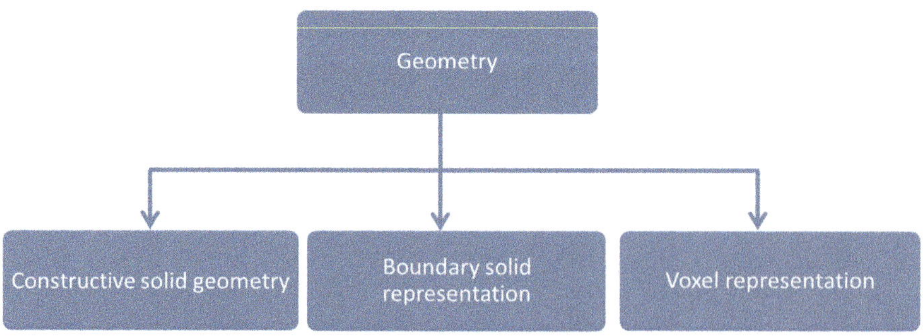

Figure 3.2. Approaches used in the design of computational anthropomorphic phantoms.

prisms, etc. Based on these shapes and using Boolean operators (or the equivalent), complex phantoms may be defined. Quadrics are surfaces defined by second order polynomial equations of three independent variables x, y, z:

$$Ax^2 + Ey^2 + Hz^2 + 2Bxy + 2Fyz + 2Cxz + 2Dx + 2Gy + 2Iz + J = 0, \quad (3.5)$$

Table 3.8. Surfaces described by quadric equations.

Quadrics	Surface equation
Real ellipsoid	$\frac{x^2}{a^2} + \frac{y^2}{b^2} + \frac{z^2}{c^2} = 1$
Hyperboloid of one sheet	$\frac{x^2}{a^2} + \frac{y^2}{b^2} - \frac{z^2}{c^2} = 1$
Hyperboloid of two sheets	$\frac{x^2}{a^2} + \frac{y^2}{b^2} - \frac{z^2}{c^2} = -1$
Real quadric cone	$\frac{x^2}{a^2} + \frac{y^2}{b^2} - \frac{z^2}{c^2} = 0$
Elliptic paraboloid	$\frac{x^2}{a^2} + \frac{y^2}{b^2} + 2z = 0$
Hyperbolic paraboloid	$\frac{x^2}{a^2} - \frac{y^2}{b^2} + 2z = 0$
Real elliptic cylinder	$\frac{x^2}{a^2} + \frac{y^2}{b^2} = 1$
Hyperbolic cylinder	$\frac{x^2}{a^2} - \frac{y^2}{b^2} = -1$
Real intersecting planes	$\frac{x^2}{a^2} - \frac{y^2}{b^2} = 0$
Parabolic cylinder	$x^2 + 2y = 0$
Real parallel planes	$x^2 = 1$
Coincident planes	$x^2 = 0$

where A, B, C, etc, are the coefficients of the equation. The quadric equations of some of the frequently used surfaces are summarised in table 3.8. The space enclosed by these surfaces is solid.

In this table a, b, and c are parameters which control the shape of the surface. For instance, in the case of an elliptic cylinder, a is the semi-major axis, while b is the semi-minor axis. In case of an ellipsoid, a, b, and c are nonzero constants, called half-axes of the ellipsoid.

When polygons are used, the object is enclosed by a set of surface polygons that enclose it interiorly. A polygon surface is specified by a set of vertex coordinates. Any plane α has a general equation of the form:

$$Ax + By + Cz + D=0, \tag{3.6}$$

where A, B, C, and D are real constants, and at least one of the coefficients A, B, C is nonzero.

To create a complex computational model of a human body or part of it, these geometrical primitives are subjected to merging or subtraction from each other using Boolean operators of union, intersection, and difference. Figure 3.3 shows the main CSG Boolean operations, applied on geometrical primitives to create a complex model.

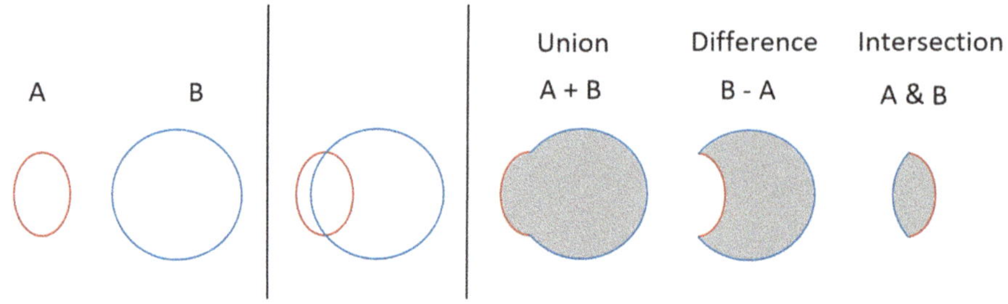

Figure 3.3. Basic Boolean operations: union, difference, and intersection, demonstrated for two shapes A and B.

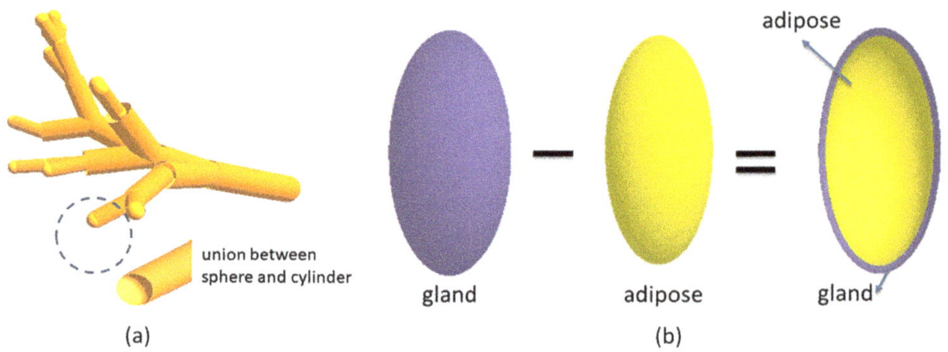

Figure 3.4. Examples for application of Boolean operations with geometrical primitives: (a) Boolean union and (b) Boolean intersection.

CSG models are widely used to represent data on acquired volume, CSG operations, and ray tracing. An example is shown in figure 3.4 for the creation of a glandular tree, which is built from cylinders terminating in spheres. In this example, the Boolean operation applied is a union between cylinders and spheres. Figure 3.4(b) demonstrates the creation of a Cooper ligament, one of the elements of the breast model discussed in chapter 5. In this case, the volume of the internal ellipsoid is assigned the properties of adipose tissue, while the Boolean operation difference results in the formation of tissue with assigned glandular characteristics.

Solid representations define solids with better accuracy and few parameters compared to the other two approaches. By using CSG, the size and the shape of the human body as well as its organs may be described easily. Examples of modelled organs in the MIRD computational phantom (figure 2.21, chapter 2) are shown in table 3.9.

For instance, the stomach wall is represented by the volume between two concentric ellipsoids. The contents are represented by the volume within the inner ellipsoid. The wall is defined by

$$\left(\frac{x-x_0}{a}\right)^2+\left(\frac{y-y_0}{b}\right)^2+\left(\frac{z-z_0}{c}\right)^2\leqslant 1 \text{ and } \left(\frac{x-x_0}{a-d}\right)^2+\left(\frac{y-y_0}{b-d}\right)^2+\left(\frac{z-z_0}{c-d}\right)^2\geqslant 1$$

with $a = 4.00$, $b = 3.00$, $c = 8.00$, $d = 0.613$, $x_0 = 8.00$, $y_0 = -4.00$, and $z_0 = 35.00$ for an adult male. The parameters are in centimetres. Figure 3.5(a) shows the result.

Table 3.9. Selected organs/tissues from the MIRD mathematical phantom (figure 2.21, chapter 2).

Organ/tissue	Mathematical description
Trunk	Solid elliptical cylinder
Head	Right elliptical cylinder topped by half an ellipsoid
Neck	Right circular cylinder
Legs	Frustums of two circular cones
Leg bones	Each leg bone is the frustum of a circular cone
Arm bones	Each arm bone is the frustum of an elliptical cone
Pelvis	A portion of the volume between two nonconcentric elliptical cylinders
Cranium	The volume between two concentric ellipsoids
Facial skeleton	A portion of the volume between two concentric elliptical cylinders; the portion of the volume that intersects the cranium and brain is excluded
Rib cage	A series of bands between two concentric, right-vertical, elliptical cylinders
Adrenals	Each adrenal is half an ellipsoid atop a kidney
Brain	An ellipsoid
Breasts	Portions of two ellipsoids attached to the trunk
Gall bladder and contents	For ages > 1 and adult male phantoms—frustum of a cone capped with a hemisphere; for the new-born phantom the bladder is cylindrical
Stomach	The stomach wall is represented by the volume between two concentric ellipsoids
The small intestine and contents	A section of an elliptical cylinder
Heart and contents	The outer surface of the heart is represented by four quarter-ellipsoids; within this space, the heart is divided into regions representing the muscular walls and the four chambers
Kidneys	Each kidney is an ellipsoid cut by a plane
Liver	The liver is defined by an elliptical cylinder cut by a plane
Lungs	Half an ellipsoid with a section removed; the section removed from the left lung is larger than that removed from the right lung because of the position of the heart
Ovaries	Each ovary is an ellipsoid
Pancreas	Half an ellipsoid with a section removed
Spleen	An ellipsoid
Testes	Ellipsoids
Thymus	An ellipsoid
Thyroid	The lobes of the thyroid lie between two concentric cylinders and are formed by a cutting surface
Urinary bladder	The bladder wall is represented by the volume between two concentric ellipsoids
Uterus	Ellipsoid cut by a plane

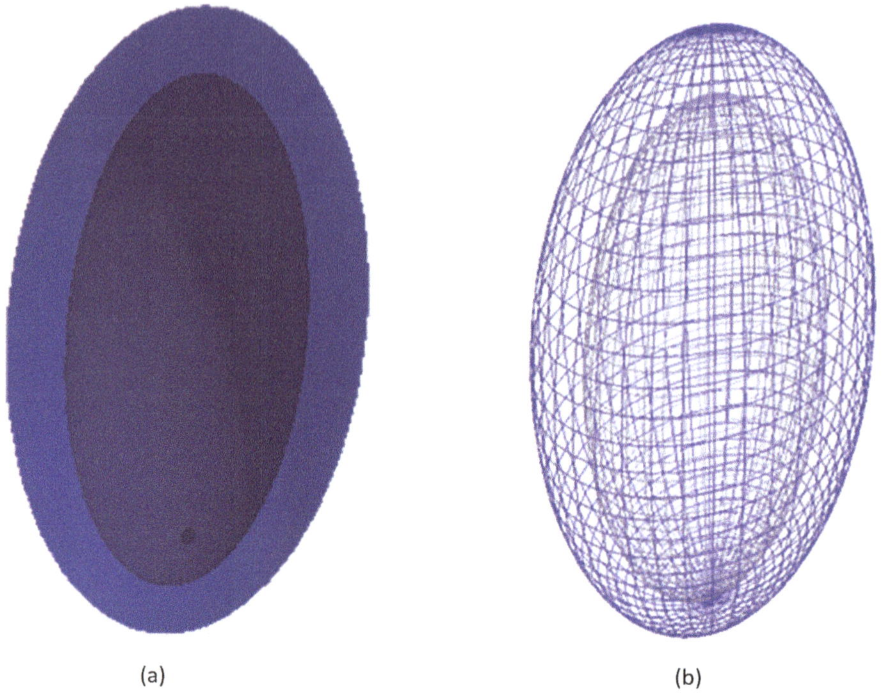

Figure 3.5. Stomach defined in a MIRD phantom: (a) CSG and (b) polygon mesh forms.

The main problem with the CSG is the robust computation of Boolean operations between solids of irregular shapes.

B-Rep models are based on the representation of surfaces. The B-Rep models consist of two types of information, geometric (NURBS curves and surfaces) and topological, allowing better links between geometrical entities. The boundary model of an object is comprised of three basic components: faces, edges, and vertices. These three components are linked in such a way as to ensure the topological consistency of the object model. The model is presented as a set of faces that bounds the physical object. The faces are regions of closed and orientable surfaces which are bounded by edges and each edge is bounded by surfaces. In addition to these three primitives, the following can be defined: loop—an ordered alternating sequence of vertices and edges; handle—a so-called through hole; a body—a so-called shell. B-Rep phantoms are usually implemented by applying the NURBS method or polygonal mesh method, shown in figure 3.5(b).

Usually objects defined in this way are polyhedral, where the object is bounded by plane faces and curved surfaces using spline surface modelling. Typically, B-Rep models are more difficult to create and require greater memory compared to CSG. However, extraction of information for rendering is quicker than for CSG.

Voxel phantoms are based on discretization of an object into tiny cubes, called voxels, with uniform characteristics. One approach to obtain voxel models is by digitising an object into voxels. Such an example is shown in figure 3.6(a), where a water CSG ellipsoid phantom is transformed into a voxel phantom with voxel sizes of 1 mm, 0.5 mm, and 0.1 mm (in each direction), shown in figures 3.6(b)–(d). Another source for voxel phantoms is 3D medical imaging techniques: CT, MRI, tomosynthesis, PET, as well as cross-sectional photographs of a cadaver through which a detailed patient-specific anatomy may be obtained. Images obtained with these modalities are represented by a matrix of pixels. A consecutive set of such images can be considered as a 3D matrix made of voxels, where each voxel belongs to a specific organ or tissue. Another example is shown in figure 3.7(a), where a model of a liver is created from patient CT data and then converted to solid-based geometry (figures 3.7(b), (c)). This phantom is defined as an assembly of 3D cubes. In general, voxel elements need not be cubes, rather they can be parallelepipeds. The properties of each voxel are stored within each voxel. These are indexes which correspond to specific tissues.

The creation of voxel phantoms starts with the data obtained using a 3D scanning technique. The projections are then used to reconstruct the volume of interest. By using various 3D reconstruction algorithms, tissues and organs are segmented in such a way that at the end numbers (HU) belonging to given organs are obtained. Finally, these numbers are all placed within a 3D matrix, comprising the voxel phantom.

The process for creating voxel phantoms is characterised by simplicity, since only trivial Boolean operations need be applied. Voxel phantoms provide very detailed

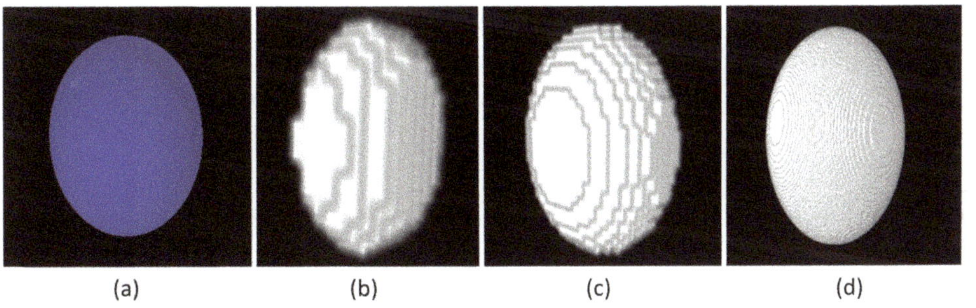

Figure 3.6. Creation of voxel phantoms: (a) a simple solid phantom and its voxel version with a voxel size of (b) 1 mm, (c) 0.5 mm, and (d) 0.1 mm in each direction.

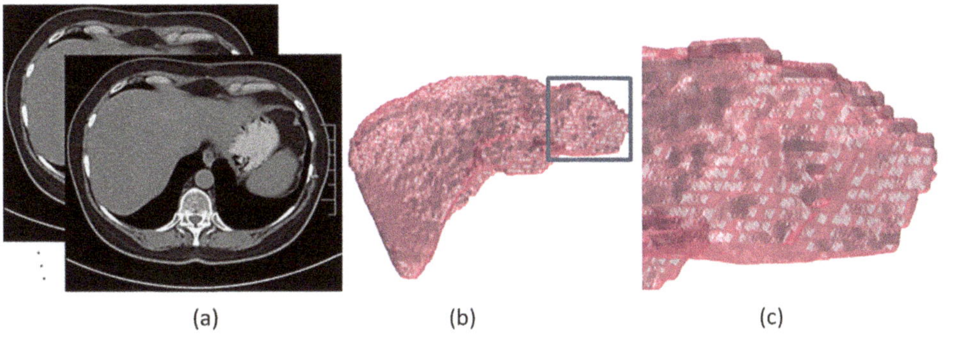

Figure 3.7. Main steps to create a voxel phantom using CT patient data: (a) CT liver images, (b) segmented liver, and (c) magnified volume.

patient-specific anatomy, which on the other hand imposes several disadvantages, amongst them the issue of storage. For example, a voxel matrix with a size of 1000 × 1000 × 1000 produces 1 billion voxels. If the data presentation of each voxel is 2 B, then the required memory to store this model will be approximately 2 GB. For a whole-body model built from high-resolution images, the memory occupied by the computational model imposes significant limitations on the ray tracing techniques. In contrast, if the voxel is large, it requires that several tissues be approximated by one. This also depends on the segmentation algorithm used. Last but not least, original voxel values are not easily rescaled to obtain scaled phantoms, since the scalability is generally limited to uniform variations in body size without changes in relative organ or body region anatomy (Lee *et al* 2010).

The variety of techniques available in computer graphics for the creation of computational objects is wide. The choice of technique depends very much on the application for which the model is designed, as well as on the computer configuration, largely the memory requirements. The main points to be considered are the reproduction of human anatomy at a sufficient level for the required application and to achieve computational efficiency, as well as assuring geometrical compatibility with the Monte Carlo code that eventually carries out the radiation transport calculation.

3.4 Test questions

- List several pros and cons of the voxel and SCG computational phantoms.
- What are the tissue-equivalent materials used to mimic the soft tissue?
- What is important when choosing a suitable tissue-equivalent material?

References

Badal A *et al* 2018 Reproducing two-dimensional mammograms with three-dimensional printed phantoms *J. Med. Imag.* **5** 033501

Carton A K *et al* 2011 Development of a physical 3D anthropomorphic breast phantom *Med. Phys.* **38** 891–896

Craft D F and Howell R M 2017 Preparation and fabrication of a full-scale, sagittal-sliced, 3D-printed, patient-specific radiotherapy phantom *J. Appl. Clin. Med. Phys.* **18** 285–92

Ferreira C C *et al* 2010 Evaluation of tissue-equivalent materials to be used as human brain tissue substitute in dosimetry for diagnostic radiology *Nucl. Instrum. Methods Phys. Res.* B **268** 2515–2521

FitzGerald , P F *et al* 2017 Liquid tissue surrogates for x-ray and CT phantom studies *Med. Phys.* **44** 6251–6260

Hazelaar C *et al* 2018 Using 3D printing techniques to create an anthropomorphic thorax phantom for medical imaging purposes *Med. Phys.* **45** 92–100

Hill R *et al* 2005 A dosimetric evaluation of water equivalent phantoms for kilovoltage x-ray beams *Phys. Med. Biol.* **50** N331–344

ICRP 2002 Basic anatomical and physiological data for use in radiological protection reference values ICRP Publication 89 *Ann. ICRP* **32** 3–4

Ikejimba L C *et al* 2017 A novel physical anthropomorphic breast phantom for 2D and 3D x-ray imaging *Med. Phys.* **44** 407–416

Ivanov D *et al* 2018 Suitability of low density materials for 3D printing of physical breast phantoms *Phys. Med. Biol.* **63** 175020

Jahnke P *et al* 2017 Radiopaque three-dimensional printing: a method to create realistic CT phantoms *Radiology* **282** 569–575

Johns H E and Cunningham J R 1983 *The Physics of Radiology* (Springfield, IL: Charles C Thomas)

Jones A K *et al* 2003 Tissue-equivalent materials for construction of tomographic dosimetry phantoms in pediatric radiology *Med. Phys.* **30** 2072–2081

Kiarashi N *et al* 2015 Development of realistic physical breast phantoms matched to virtual breast phantoms based on human subject data *Med. Phys.* **42** 4116–4126

Lee C *et al* 2010 The UF family of reference hybrid phantoms for computational radiation dosimetry *Phys. Med. Biol.* **55** 339–363

Pantelis E *et al* 2004 Polymer gel water equivalence and relative energy response with emphasis on low photon energy dosimetry in brachytherapy *Phys. Med. Biol.* **49** 3495–3514

Qi Z *et al* 2010 Quantitative imaging of electron density and effective atomic number using phase contrast CT *Phys. Med. Biol.* **55** 2669–2677

White D R *et al* 1977 Epoxy resin based tissue substitutes *Br. J. Radiol.* **50** 814–821

White D R *et al* 1978 Epoxy-resin-based tissue substitutes *Med. Phys.* **5** 453–454

Winslow J F *et al* 2009 Construction of anthropomorphic phantoms for use in dosimetry studies *J. Appl. Clin. Med. Phys.* **10** 2986

IOP Publishing

Anthropomorphic Phantoms in Image Quality and Patient Dose Optimization

A EUTEMPE Network book

Kristina Bliznakova, Ivan Buliev and Zhivko Bliznakov

Chapter 4

Modelling and simulation of x-ray imaging techniques

4.1 Introduction

Computer modelling and simulation is extensively used in the development of new diagnostic imaging and radiotherapy modalities. The use of such tools is becoming of critical importance in designing, testing, and predicting the performance of forthcoming x-ray imaging systems, prior to their manufacture and before being accepted for clinical use. One such example is the advanced applications of full-field digital mammography systems, especially those that provide three-dimensional images of the breast, such as digital breast tomosynthesis, cone-beam breast CT, and breast CT. To provide a faithful simulation of these systems, all components must be accurately modelled, followed by verification through experimental trials.

The main components of x-ray imaging simulations are image acquisition, during which the projection data are acquired, and tomographic reconstruction (in the case of 3D imaging). The first step requires modelling of the acquisition geometry, the patient (phantom), and a computer code simulating the radiation transport in the model. For instance, in the case of breast imaging, modelling of the following components of the x-ray imaging system are required: x-ray spectra, detector, acquisition geometry, and the breast itself, and a computer code to simulate the radiation transport in the model, resulting in the generation of projection mammography images and evaluation of the dose absorbed in the breast. The precise modelling of all these components will ensure results approaching what would be obtained in practice.

In this chapter, attention is paid to the computer modelling of x-ray based imaging techniques and their use in simulations in standard acquisition geometries in diagnostic radiology, as well as for studying novel x-ray imaging techniques. The aim is to learn how to model each part of the x-ray system, the patient, and the image formation. A short overview of the main components of an x-ray imaging system is provided as well. At the end of the sections, exercises are provided. They will be carried out using the in-house developed software applications, described in the following section.

4.2 Overview of the *XRAYImagingSimulator* software tool

XRAYImagingSimulator is a complete software package dedicated to x-ray imaging research studies (Bliznakova *et al* 2010). With its assistance, one can design computational phantoms from *simple 3D geometrical primitives* (chapter 3) or to import patient or other data scanned in imaging modalities in the form of *voxelised phantoms* (chapter 3). The *XRAYImagingSimulator* application comprises a set of modules that address specific modelling issues. The platform that is schematically presented in figure 4.1 comprises four basic software modules: (a) the phantom modelling module—used to generate 3D phantoms composed of voxels or solid geometrical primitives; (b) the geometry modelling module—used to model the acquisition geometry: isocentric rotational geometry, mammography, cone-beam CT, or tomosynthesis; (c) the image formation module—used to model the x-ray transport and generate synthetic projection images; and (d) the visualisation module—a set of utilities to visualise 3D software phantoms and projection images.

The executable for Windows, with relevant help and demo files, will be available for the participants in the EUTEMPE course. To install it, use the *SoftwareRequirements.txt* file, which contains detailed information about the recommended PC configuration.

The main window of the application is depicted in figure 4.2. In brief, the *XRAYImagingSimulator* allows modelling of:

Phantoms: composed of geometrical primitives and voxel-based phantoms.

Acquisition geometries: radiography, standard mammography, tomosynthesis, CT, and cone-beam and helical CT.

Beam type: monochromatic and polychromatic beams.

Detector: models of phosphor screens for phosphor-based detectors (e.g. CsI, Gd_2O_2S), coupled with charge-coupled devices, and ideal detectors.

Transport simulation: the energy reaching the detector pixel is calculated using Lambert–Beer's law.

Visualisation of images and 3D computational models.

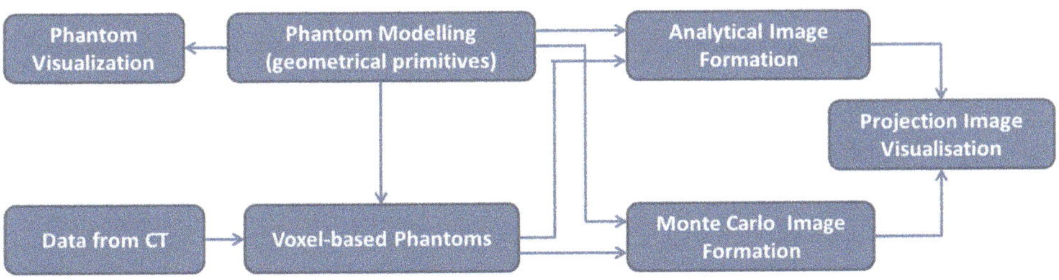

Figure 4.1. The main modules in the software application.

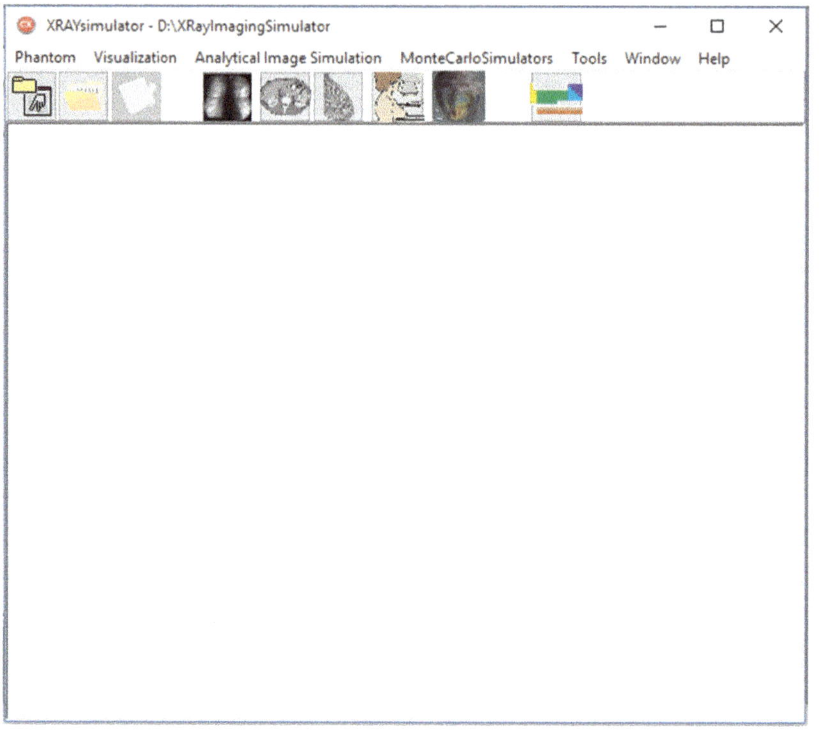

Figure 4.2. The main window of the *XRAYImagingSimulator*.

Figure 4.3 shows an example of use of this application. Two computational phantoms, voxel- and solid-based, were created (figures 4.3(a), (b)) and their simulated x-ray projection images were produced (figures 4.3(c), (d)).

4.3 Creating solid-based geometry anthropomorphic phantoms

4.3.1 Tutorial for the creation of CSG phantoms

The aim of this tutorial is to review the main steps in the design and the creation of solid-based anthropomorphic phantoms. A description of the software functionalities needed to create the computational phantoms is provided at the following link: http://tuv.eutempe-net.eu/wp-content/uploads/Education-materials/Videos/CreateSolidPhantom%206_4.mp4.

4.3.1.1 Phantom modelling

The creation of computational phantoms starts with the *Phantom* menu, which is shown in figure 4.4. The *New* and *Open Phantom* options are dedicated to phantoms built as a synthesis of simple uniform geometrical shapes (CSG-based) with user-defined dimensions, location, orientation, and composition.

New Phantom is an option which applies to the creation of new phantoms, composed of 3D geometrical primitives. These are characterised by an object type and property, and these properties are selected using the combo boxes, shown in

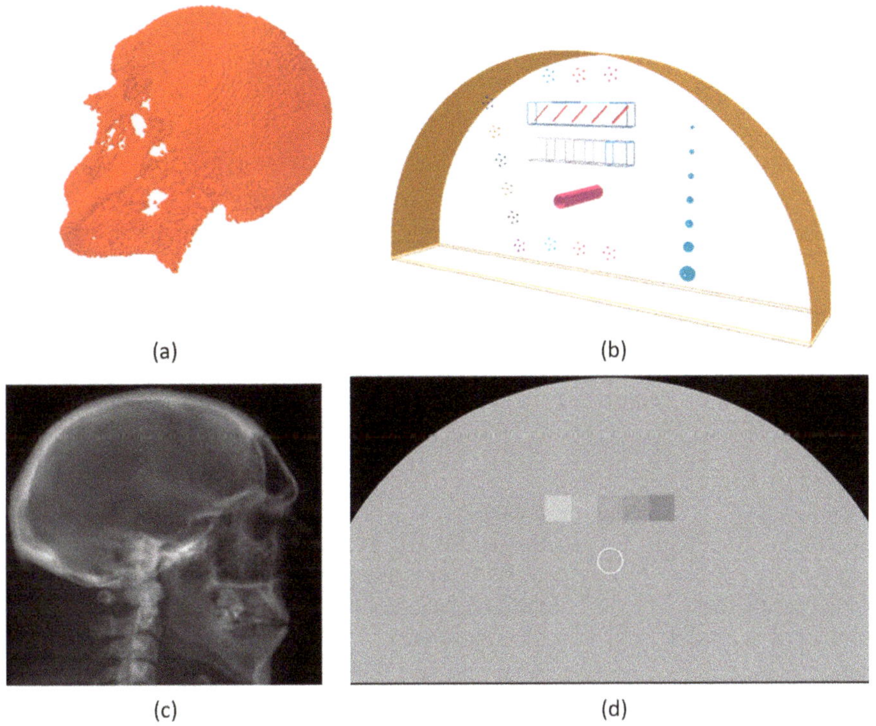

Figure 4.3. Applications of the *XRAYImagingSimulator*: (a) voxel and (b) CSG computational phantoms and their projection images in (c) and (d), respectively.

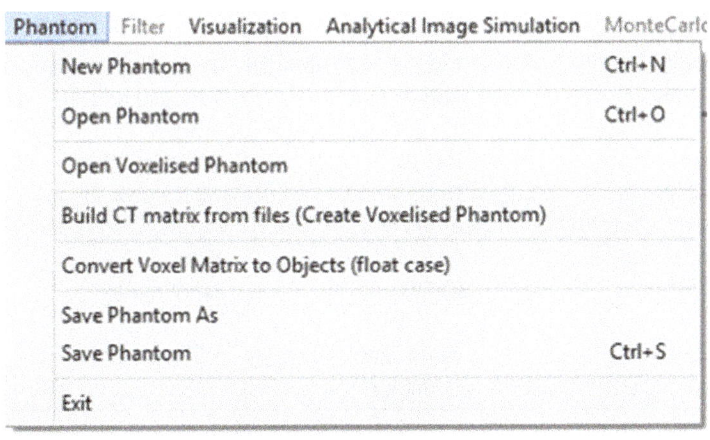

Figure 4.4. *Phantom* module option.

figures 4.5(a),(b). When constructing phantoms, the *Object Property* must be selected from the corresponding combo box. If the simulation purpose is to use analytical image formation, then the detector property is not applicable for this case. The *Object Type* specifies the geometrical primitive to be used in the phantoms. The modelled geometrical primitives are listed in table 4.1.

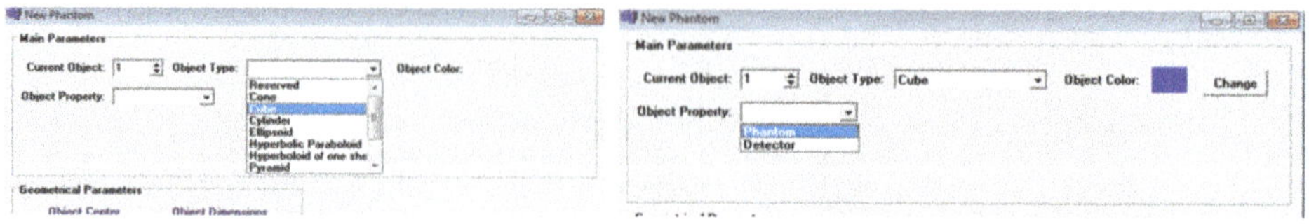

(a) (b)

Figure 4.5. The *New Phantom* option is used to create a new computational phantom.

Table 4.1. Description of the object type and composition list used in the description of each object.

Object	Object type	Composition list		
Cone	Phantom	Water	Soft tissue ICRU-44	Blood
Parallelepiped	Detector	Plexiglas	Ideal air	SiO_2
Cylinder		Adipose ICRU-44	SrFBr	Silicon gel
Ellipsoid		$CaWO_4$ (phosphor)	Paraffin wax	Epoxy resin
Hyperbolic paraboloid		Gd_2O_2S (phosphor)	Cerrobend	Nylon-6,6
Hyperboloid of one sheet		CsI (phosphor)	Rad material	Water vapour
Pyramid		Bone compact ICRU-44	Brass	Ethanol
Semi-ellipsoid		Ideal detector	Concrete	Ethanol vapour
Elliptic paraboloid		Muscle ICRU-44	Gravitech 8.15	Eye lens
Semi-cylinder		Air	Gravitech – 11	Polyethylene
Oblique cylinder		Bone cortical ICRU-44	Ecomass	Povidone-iodine
Astroidal ellipsoid		Lung tissue—inflated	Technon poly kit	Al_2O_3
Custom semi-cylinder		Polystyrene	75% gland, 25% fat	Hydroxyapatite
Elliptic cylinder		BaFBr	Br50%–50%	Clear resin
Custom semi-ellipsoid		Iroco wood	50% gland, 50% fat	
Hyperboloid of one sheet		Gland	Skin	
1 Hyperboloid of one sheet		$CaCO_3$	Gray matter	
2 Hyperboloid of one sheet		Brain, grey/white matter	White matter	

Further, the dimensions and the absorption characteristics of each 3D geometrical primitive which is part of the computational phantom are specified in the corresponding edit boxes of the *Geometrical Parameters* panel, as shown in figure 4.6. The dimensions and the position in 3D space, described by a Cartesian coordinate system, are specified in millimetres. The rotation of the geometrical primitive is

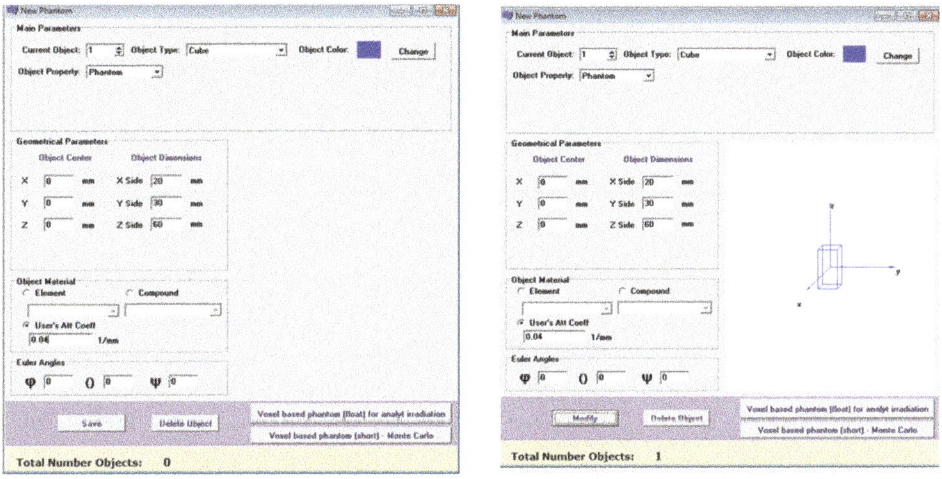

(a) (b)

Figure 4.6. Main phases in the creation of a phantom from solid objects.

implemented by means of Euler angles (https://en.wikipedia.org/wiki/Euler_angles). These angles are examined in detail in section 4.3.1.4 with a specific example. The entered parameters are stored by clicking the button *SAVE* (figure 4.6(a)). The absorption characteristics of the 3D geometrical primitive are implemented either by a linear attenuation coefficient (μ), specified in mm^{-1} or by choosing any material from the combo boxes *Element* or *Compound*. The *Element* combo box contains the elements from the periodic table, while the *Compound* combo box contains more than 50 modelled phantom materials, which are listed in the third column of table 4.1. The *User's Att Coeff* edit box gives the possibility to directly specify the linear attenuation coefficients of the object. This may be used when a phantom material is not in the database of the program. An example, shown in the case of water in figure 4.7, reveals the consequence of the actions to obtain μ and to use that value in the program. The mass attenuation coefficients for any element, compound, or mixture may be obtained with the XCOM program http://www.nist.gov/pml/data/xcom/. The linear attenuation coefficient, μ, is then obtained by multiplying the mass attenuation coefficient with the physical density of the material and this value is used in the *User's Att Coeff* edit box.

An example of a phantom containing one parallelepiped object with (X, Y, Z) dimensions 20 mm × 30 mm × 60 mm is shown in figure 4.6(b).

The values in all these edit and combo boxes are obligatory for the full characterisation of any computational phantom. The number of objects contained in a computational phantom are displayed in *Total Number Objects*. A new object may be added by increasing the number in the *Current Object* spin button. To add a new object, the same procedure as described above is followed. It is necessary to confirm new/modified object data with the button *Save/Modify*. The use of *Delete Object* deletes an object from the computational phantom.

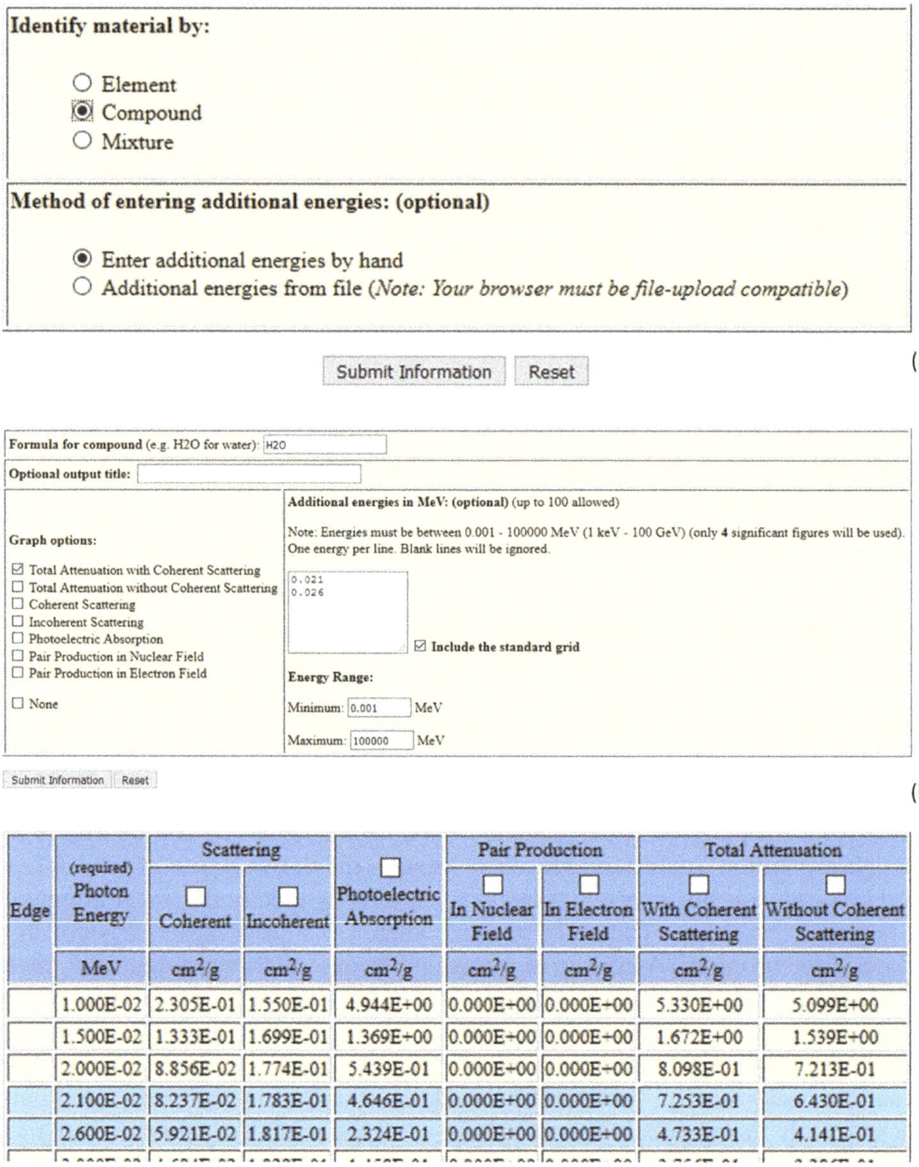

Edge	(required) Photon Energy	Scattering		Photoelectric Absorption	Pair Production		Total Attenuation	
		Coherent	Incoherent		In Nuclear Field	In Electron Field	With Coherent Scattering	Without Coherent Scattering
	MeV	cm²/g	cm²/g	cm²/g	cm²/g	cm²/g	cm²/g	cm²/g
	1.000E-02	2.305E-01	1.550E-01	4.944E+00	0.000E+00	0.000E+00	5.330E+00	5.099E+00
	1.500E-02	1.333E-01	1.699E-01	1.369E+00	0.000E+00	0.000E+00	1.672E+00	1.539E+00
	2.000E-02	8.856E-02	1.774E-01	5.439E-01	0.000E+00	0.000E+00	8.098E-01	7.213E-01
	2.100E-02	8.237E-02	1.783E-01	4.646E-01	0.000E+00	0.000E+00	7.253E-01	6.430E-01
	2.600E-02	5.921E-02	1.817E-01	2.324E-01	0.000E+00	0.000E+00	4.733E-01	4.141E-01

(c)

Figure 4.7. Obtaining linear attenuation coefficients μ using the XCOM program: (a) a panel for selecting the way the material is defined, (b) material definition with additional energies specified by the user, and (c) the calculated mass absorption coefficients.

The computational phantom data are saved in a file by choosing the *Save Phantom* or *Save Phantom As* from the *Phantom* menu. The file is stored in **.phn* format and can be opened by any text editor. It contains all the information for each object starting from the object type (represented as a number), its coordinates, dimensions, object property, composition, Euler angles, object colour, etc. An

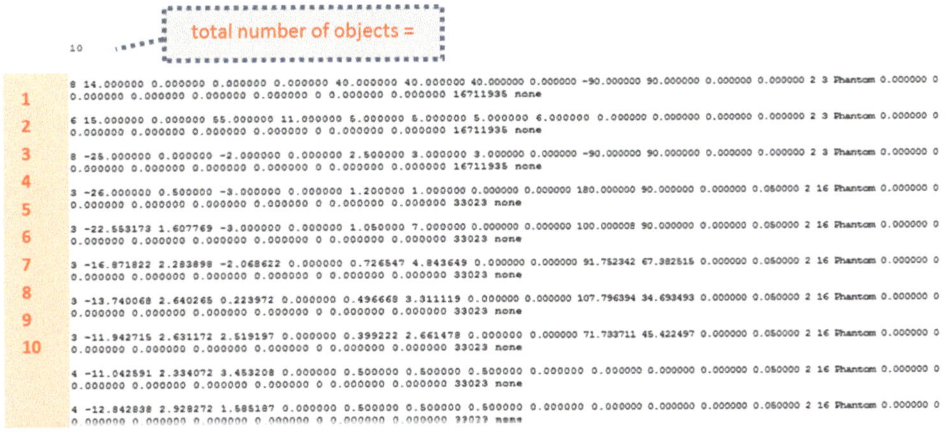


```
10
8 14.000000 0.000000 0.000000 0.000000 40.000000 40.000000 40.000000 0.000000 -90.000000 90.000000 0.000000 0.000000 2 3 Phantom 0.000000 0
0.000000 0.000000 0.000000 0.000000 0 0.000000 0.000000 16711935 none
6 15.000000 0.000000 55.000000 11.000000 5.000000 5.000000 5.000000 6.000000 0.000000 0.000000 0.000000 0.000000 2 3 Phantom 0.000000 0
0.000000 0.000000 0.000000 0.000000 0 0.000000 0.000000 16711935 none
8 -25.000000 0.000000 -2.000000 0.000000 2.500000 3.000000 3.000000 0.000000 -90.000000 90.000000 0.000000 0.000000 2 3 Phantom 0.000000 0
0.000000 0.000000 0.000000 0.000000 0 0.000000 0.000000 16711935 none
3 -26.000000 0.500000 -3.000000 0.000000 1.200000 1.000000 0.000000 0.000000 180.000000 90.000000 0.000000 0.050000 2 16 Phantom 0.000000 0
0.000000 0.000000 0.000000 0.000000 0 0.000000 0.000000 33023 none
3 -22.553173 1.607769 -3.000000 0.000000 1.050000 7.000000 0.000000 0.000000 100.000008 90.000000 0.000000 0.050000 2 16 Phantom 0.000000 0
0.000000 0.000000 0.000000 0.000000 0 0.000000 0.000000 33023 none
3 -16.871822 2.283898 -2.068622 0.000000 0.726547 4.843649 0.000000 0.000000 91.752342 67.382515 0.000000 0.050000 2 16 Phantom 0.000000 0
0.000000 0.000000 0.000000 0.000000 0 0.000000 0.000000 33023 none
3 -13.740068 2.640265 0.223972 0.000000 0.496668 3.311119 0.000000 0.000000 107.796394 34.693493 0.000000 0.050000 2 16 Phantom 0.000000 0
0.000000 0.000000 0.000000 0.000000 0 0.000000 0.000000 33023 none
3 -11.942715 2.631172 2.519197 0.000000 0.399222 2.661478 0.000000 0.000000 71.733711 45.422497 0.000000 0.050000 2 16 Phantom 0.000000 0
0.000000 0.000000 0.000000 0.000000 0 0.000000 0.000000 33023 none
4 -11.042591 2.334072 3.453208 0.000000 0.500000 0.500000 0.500000 0.000000 0.000000 0.000000 0.000000 0.050000 2 16 Phantom 0.000000 0
0.000000 0.000000 0.000000 0.000000 0 0.000000 0.000000 33023 none
4 -12.842838 2.928272 1.585187 0.000000 0.500000 0.500000 0.500000 0.000000 0.000000 0.000000 0.000000 0.050000 2 16 Phantom 0.000000 0
0.000000 0.000000 0.000000 0.000000 0 0.000000 0.000000 33023 none
```

Figure 4.8. The content of a *.*phn* file. The content of this file may be viewed with any text editor.

example is shown in figure 4.8 for a computational phantom composed of ten objects. Phantoms are stored in the */Phantoms* directory.

All CSG computational phantoms are saved as *.*phn* files. These files are opened using the *Open Phantom* option from the *Phantom* menu (figure 4.4). An example is presented in figure 4.9, showing a cylindrical object that is a part of the computational CIRS 0110 phantom (the physical version of which is shown in figure 2.11, chapter 2), consisting of 93 geometrical objects (the number is shown against the label *Total Number Objects*). The cylinder is characterised by its 3D location (*X*, *Y*, *Z*) and dimensions: radius and height, as well as attenuation coefficient, and Euler angles. The characteristics of the modelled 3D primitives can be reviewed by changing the horizontal offset of the frame of the spin box relative to the visual region shown in *Current Object*.

Changes in the parameters of the phantom are confirmed using the *Modify* button.

4.3.1.2 Phantom visualisation

Three-dimensional computational phantoms, composed of solid objects (i.e. the files with the extension *.*phn*) are visualised by choosing one of the two options in the *Visualization* menu. Two different visualisation modules that differ in their visualisation implementation method may be used (figure 4.10).

Examples for visualisation of two CSG phantoms are demonstrated in figures 4.11(a), (b) (breast phantom) and 4.12(a), (b) (CIRS phantom) using these two approaches.

4.3.1.3 The creation of basic solid geometrical shapes

Phantoms based on 3D geometrical primitives are built using the utilities in the *Phantom* modelling module. The implementations of the different geometrical primitives are shown in table 4.2, where the column 'Object type' corresponds to the item chosen from the combo box *Object Type*, while in column 'Example

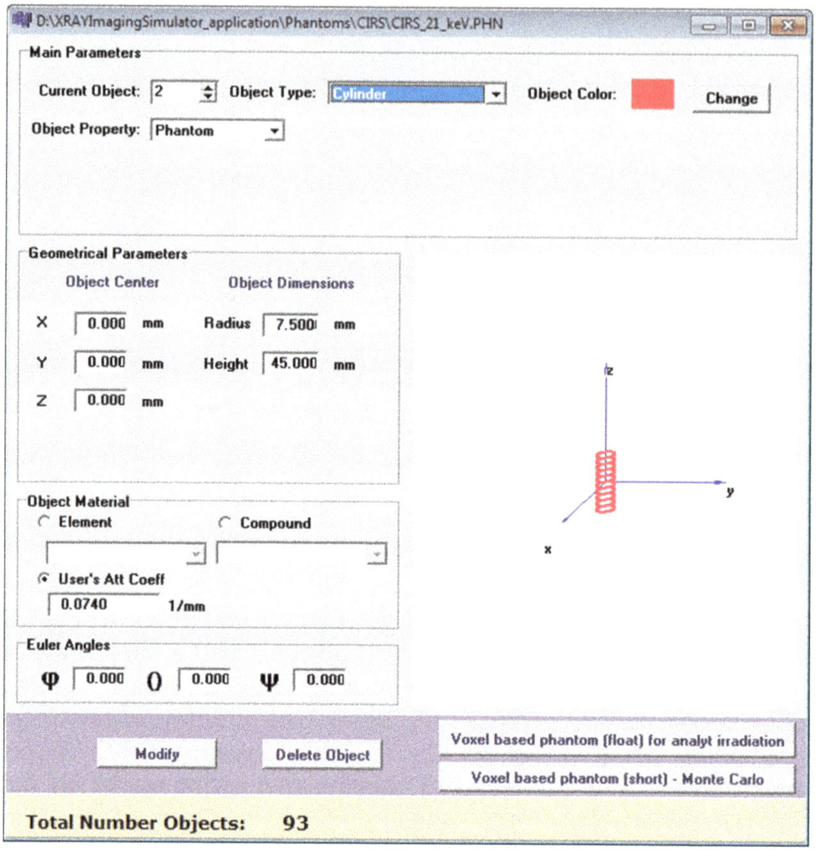

Figure 4.9. CIRS 0110 loaded using the *Open Phantom* option.

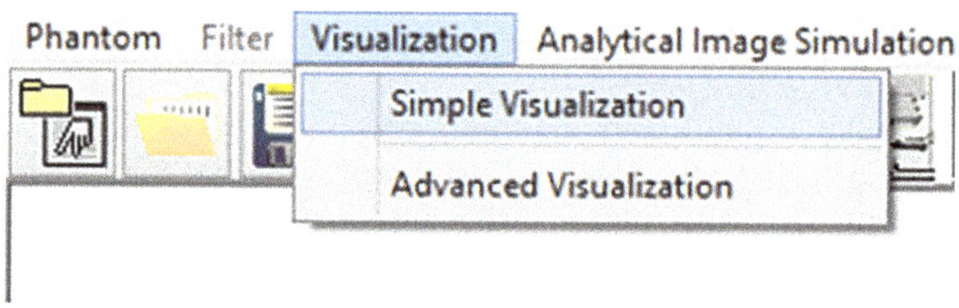

Figure 4.10. Phantom visualisation menu.

settings' are given example values which correspond to the geometrical parameters of the primitive (figure 4.6). The last column 'Implementation' shows the generated primitive using the second visualisation implementation (figure 4.10).

4.3.1.4 Euler angles

In computational physics, the Euler angles are often preferred to define object rotation. The rotation given by Euler angles (φ, θ, ψ) splits the complete rotation of a

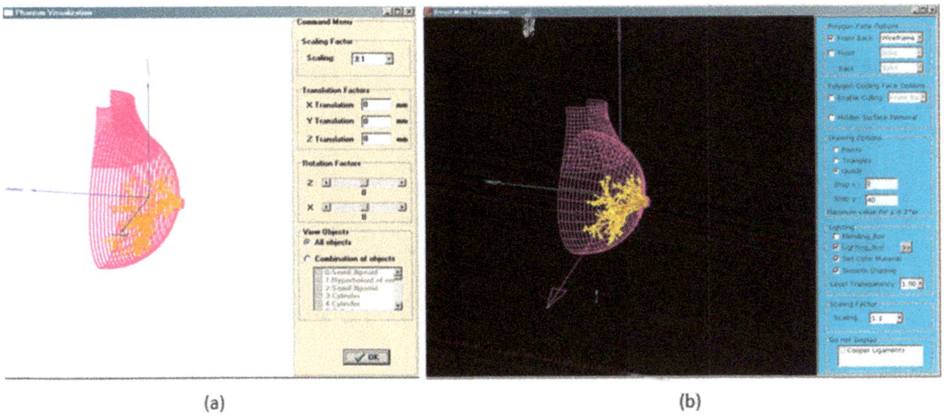

Figure 4.11. Visualisation of a breast model using (a) simple and (b) advanced visualisation techniques.

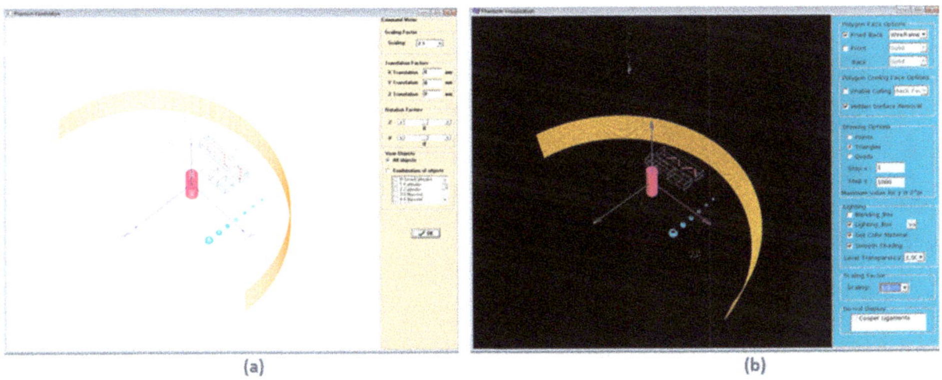

Figure 4.12. Visualisation of CIRS Model 0110 using (a) simple and (b) advanced visualisation techniques.

Cartesian coordinate system into three simpler rotations about the axes of this system: the first rotation is by an angle φ about the z-axis, the second is by an angle θ about the rotated x-axis, denoted as x', and the third is by an angle ψ about the new z-axis, denoted as z'' (figure 4.13). This is the so-called main convention.

The coordinates of a point P, rotated at Euler angles (φ, θ, ψ), in the global coordinate system $P'''(x''', y''', z''')$ with respect to the initial $P(x, y, z)$, are defined as:

$$P'''(x''', y''', z''') = R_{\text{euler}*}P(x, y, z). \tag{4.1}$$

The above equation presented in a matrix form is:

Table 4.2. Implementation of basic geometrical shapes.

Object type	Example settings	Implementation
Ellipsoid	$X = 0$, X-axis = 10 mm $Y = 0$, Y-axis = 20 mm $Z = 0$, Z-axis = 30 mm $\phi = 0°$, $\psi = 0°$, $\theta = 0°$, 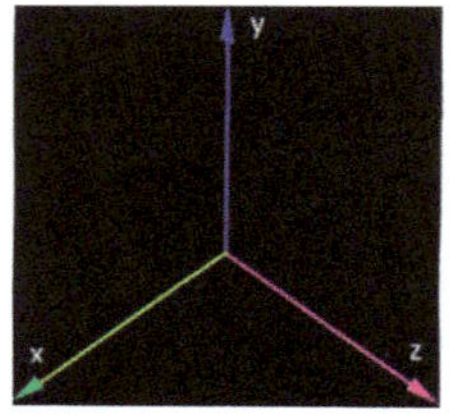 where the X-axis, Y-axis, and Z-axis are the semi-axes and correspond to the coefficients a, b, and c defined in chapter 3.	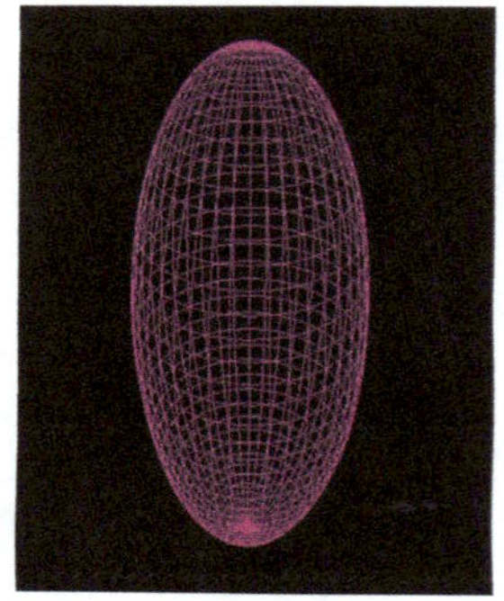
Semi-ellipsoid (implemented for $z > 0$)	$X = 0$, X-axis = 10 mm $Y = 0$, Y-axis = 20 mm $Z = 0$, Z-axis = 30 mm $\phi = 0°$, $\psi = 0°$, $\theta = 0°$, where the X-axis, Y-axis, and Z-axis are the semi-axes and correspond to the coefficients a, b, and c defined in table 3.8, chapter 3.	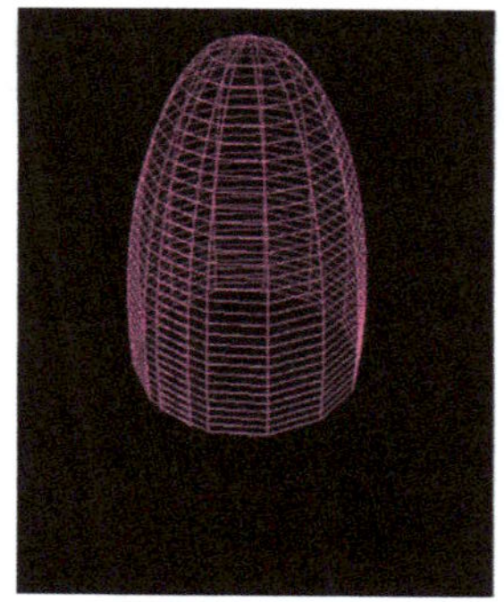

(*Continued*)

Custom semi-ellipsoid	$X = 0$, X-axis = 30 mm $Y = 0$, Y-axis = 50 mm $Z = 0$, Z-axis = 60 mm cutting angle = 4°, where the X-axis, Y-axis, and Z-axis are the semi-axes and correspond to the coefficients a, b, and c defined in table 3.8, chapter 3.	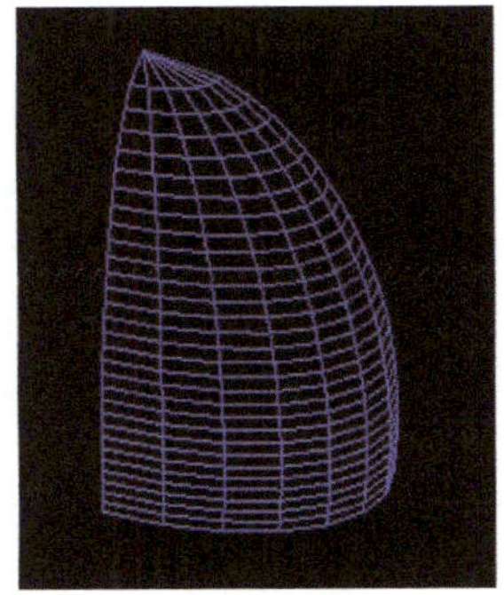
Elliptic cylinder	$X = 0$, X-axis = 20 mm $Y = 0$, Y-axis = 30 mm $Z = 0$, height = 60 mm $\phi = 0°$, $\psi = 0°$, $\theta = 0°$, where the X-axis and Y-axis are the semi-axes and correspond to the coefficients a, b from table 3.8, chapter 3.	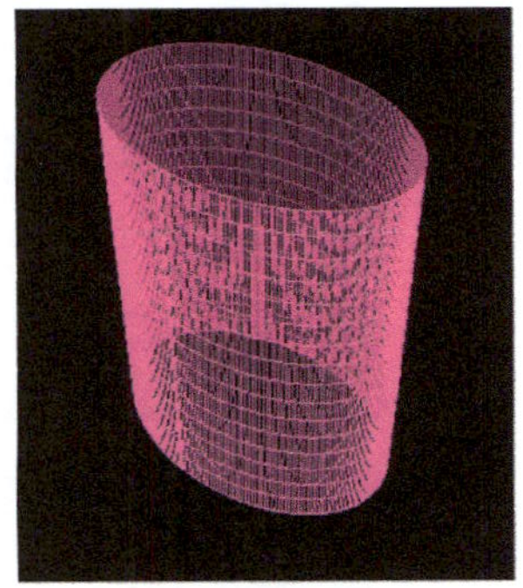
Right cylinder	$X = 0$, radius = 30 mm $Y = 0$, height = 60 mm $Z = 0$, $\phi = 0°$, $\psi = 0°$, $\theta = 0°$, where the radius is the X-axis and Y-axis, which are the semi-axes (coefficients a, b) from table 3.8, chapter 3.	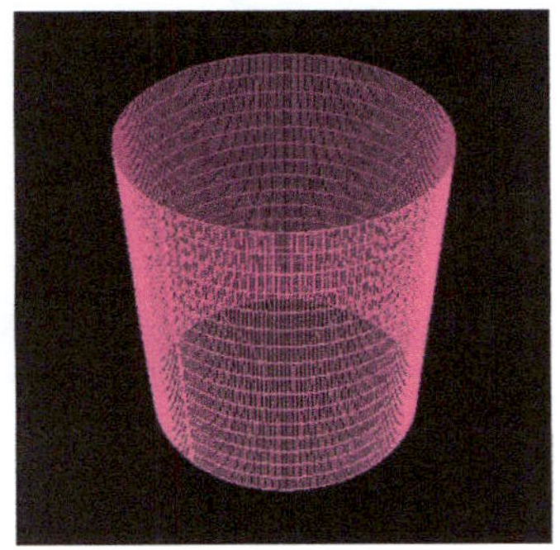

Semi-cylinder (implemented for $x > 0$)	$X = 0$, radius = 30 mm $Y = 0$, height = 60 mm $Z = 0$, $\phi = 0°$, $\psi = 0°$, $\theta = 0°$, where the radius is the X-axis and Y-axis which are the semi-axes (coefficients a, b) from table 3.8, chapter 3.	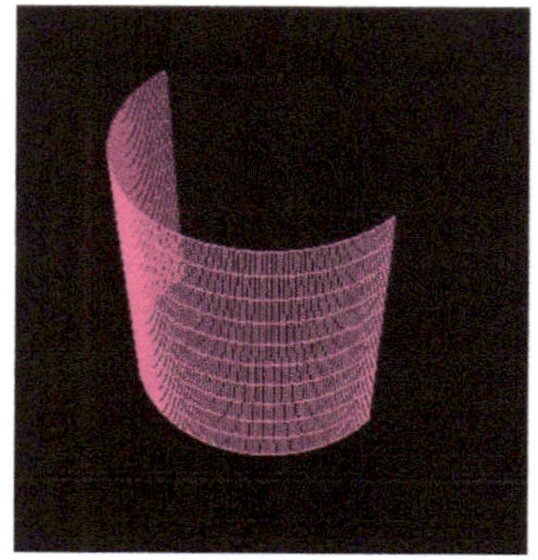
Oblique cylinder	$X = 0$, radius = 30 mm $Y = 0$, height = 60 mm $Z = 0$, angle = 50° $\phi = 0°$, $\psi = 0°$, $\theta = 0°$	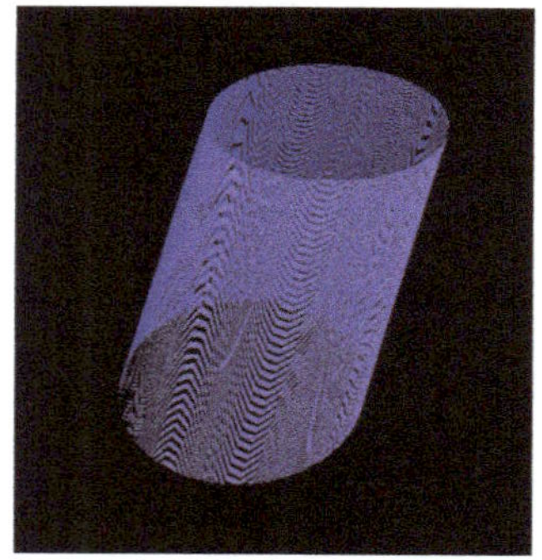
Custom cylinder	$X = 0$, radius = 60 mm $Y = 0$, height = 100 mm $Z = 0$, angle = 3° $\phi = 0°$, $\psi = 0°$, $\theta = 0°$	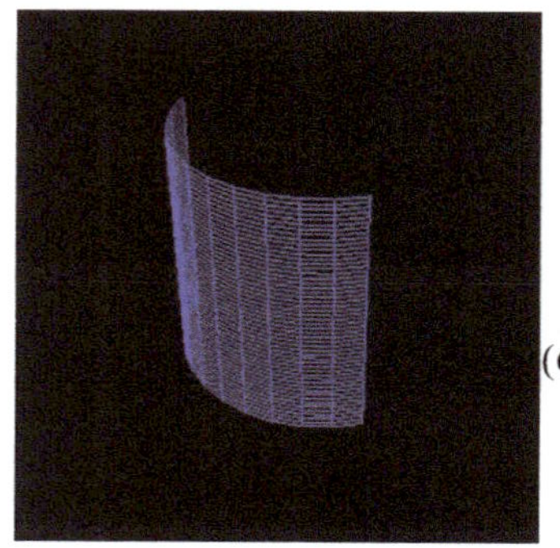

(Continued)

Right cone	$X = 0$, X-axis $= 30$ mm $Y = 0$, Y-axis $= 30$ mm $Z = 0$, Z-axis $= 80$ mm $\phi = 0°$, $\psi = 0°$, $\theta = 0°$,	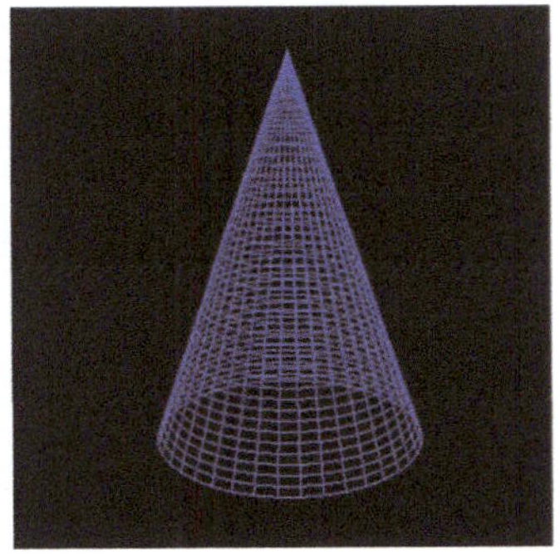
Elliptic cone	$X = 0$, X-axis $= 30$ mm $Y = 0$, Y-axis $= 60$ mm $Z = 0$, Z-axis $= 80$ mm $\phi = 0°$, $\psi = 0°$, $\theta = 0°$	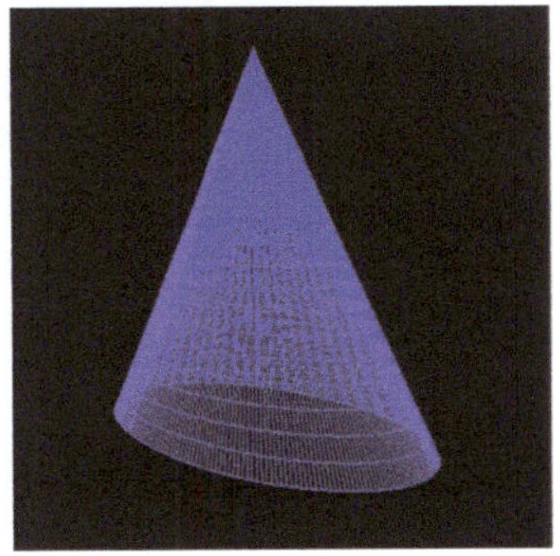
Hyperboloid of one sheet	$X = 0$, X-axis $= 4$ mm $Y = 0$, Y-axis $= 4$ mm $Z = 0$, Z-axis $= 6$ mm offset = 5, c_param = 5 $\phi = 0°$, $\psi = 0°$, $\theta = 0°$	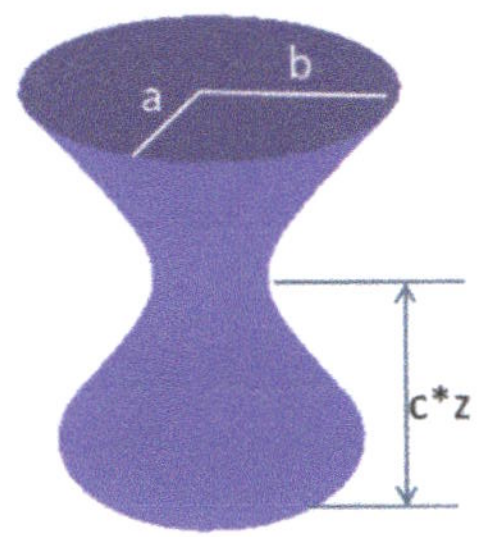

Customised hyperboloid of one sheet	$X = 0$, X-axis = 4 mm $Y = 0$, Y-axis = 4 mm $Z = 30$, Z-axis = 6 mm offset = 5, c_param = 5 $\phi = 0°$, $\psi = 0°$, $\theta = 0°$	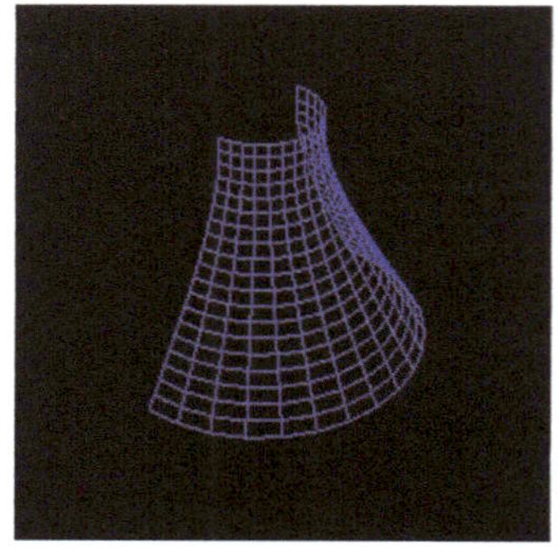
Hyperboloid of one sheets (half)	$X = 0$, X-axis = 4 mm $Y = 0$, Y-axis = 4 mm $Z = 0$, Z-axis = 6 mm offset = 5, c_param = 6 $\phi = 0°$, $\psi = 0°$, $\theta = 0°$	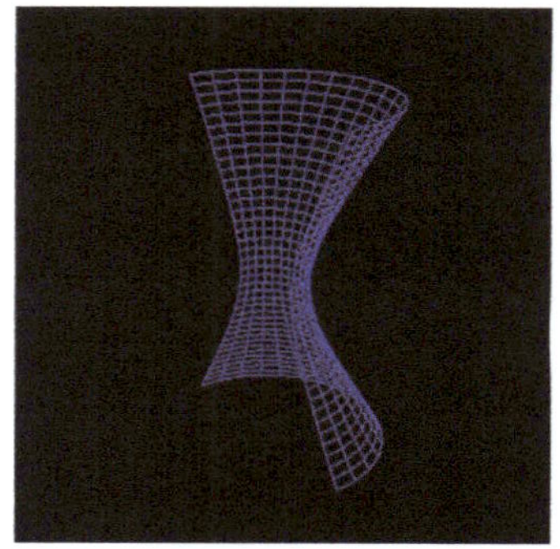
Hyperboloid of one sheets (half)	$X = 0$, X-axis = 4 mm $Y = 0$, Y-axis = 4 mm $Z = 0$, Z-axis = 6 mm offset = 5, c_param = 6 $\phi = 0°$, $\psi = 0°$, $\theta = 0°$	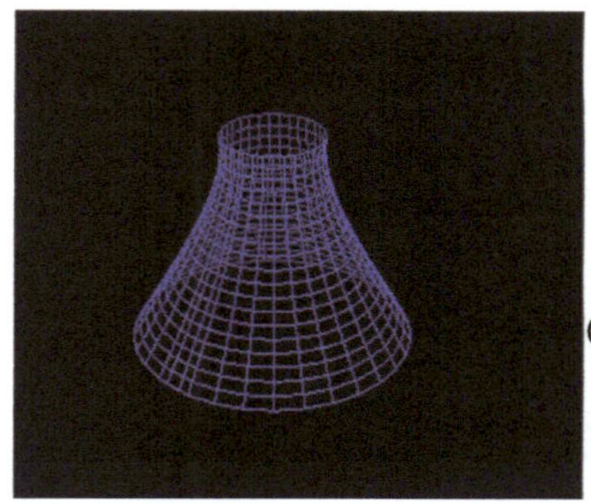

(*Continued*)

Elliptic paraboloid	$X = 0$, X-axis = 10 mm $Y = 0$, Y-axis = 10 mm $Z = 0$, Z-axis = 80 mm $\phi = 0°$, $\psi = 0°$, $\theta = 0°$	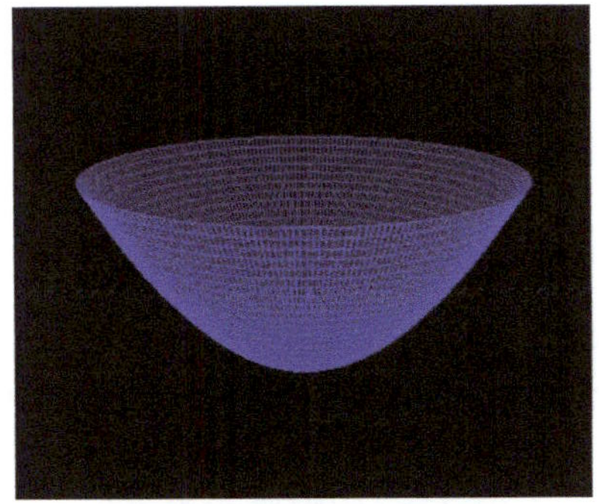
Astroidal ellipsoid	$X = 0$, X-axis = 10 mm $Y = 0$, Y-axis = 20 mm $Z = 0$, Z-axis = 30 mm $\phi = 0°$, $\psi = 0°$, $\theta = 0°$	
Cuboid	$X = 0$, X-side = 60 mm $Y = 0$, Y-side = 20 mm $Z = 0$, Z-side = 90 mm $\phi = 45°$, $\psi = 30°$, $\theta = 0°$	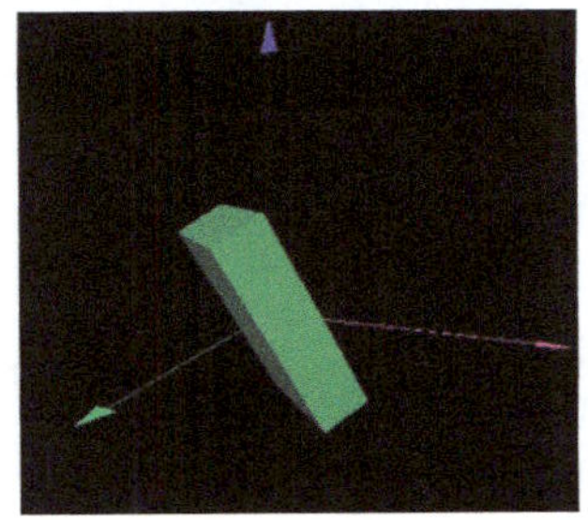

Prism	$X = 0$, X-side = 40 mm $Y = 0$, Y-side = 40 mm $Z = 0$, Z-side = 80 mm angles = 10° $\phi = 0°$, $\psi = 0°$, $\theta = 0°$	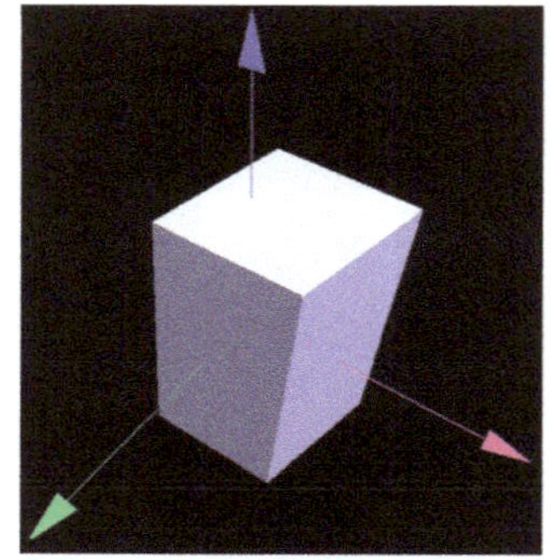
Pyramid	$X = 0$, X-side = 40 mm $Y = 0$, Y-side = 40 mm $Z = 0$, height = 100 mm $\phi = 0°$, $\psi = 0°$, $\theta = 0°$	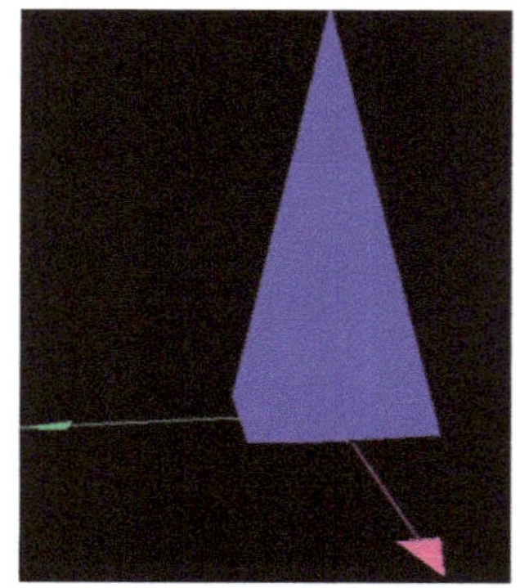

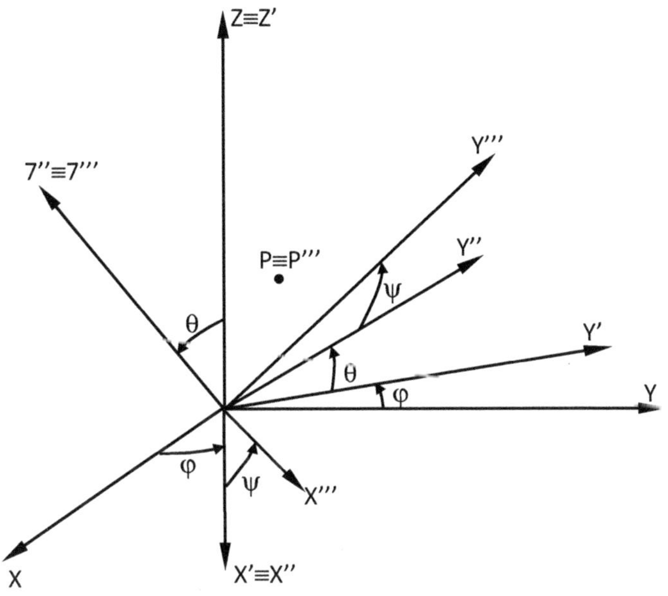

Figure 4.13. Illustration of the main convention system.

$$\begin{bmatrix} x''' \\ y''' \\ z''' \end{bmatrix} = \begin{bmatrix} \cos\psi^*\cos\phi - \sin\psi^*\cos\theta^*\sin\phi & -\sin\psi^*\cos\phi - \cos\psi^*\cos\theta^*\sin\phi & \sin\theta^*\sin\phi \\ \cos\psi^*\sin\phi + \sin\psi^*\cos\theta^*\cos\phi & -\sin\psi^*\sin\phi + \cos\psi^*\cos\theta^*\cos\phi & -\sin\theta^*\cos\phi \\ \sin\psi^*\sin\theta & \cos\psi^*\sin\theta & \cos\theta \end{bmatrix} * \begin{bmatrix} x \\ y \\ z \end{bmatrix}. \quad (4.2)$$

Using the above transformation, every three-dimensional vector can be rotated at Euler angles (φ, θ, ψ), and receive its new orientation parameters in the global system. An example for a cylindrical object, rotated about x at −45° is shown in figure 4.14.

4.3.2 Self-training exercises

Exercise 1: Computational phantoms based on geometrical primitives.
The aim of this self-training exercise is to create a computational phantom suitable for studies with dual-energy imaging. The computational phantom, composed of four objects, is modelled as a parallelepiped with three water spheres placed at different z positions with no overlap between the spheres. The phantom with example parameters described in table 4.3 is shown in figure 4.15.

Tasks: (a) create the phantom; (b) visualise it; and (c) save the phantom as *paraffin_phantom.phn*.

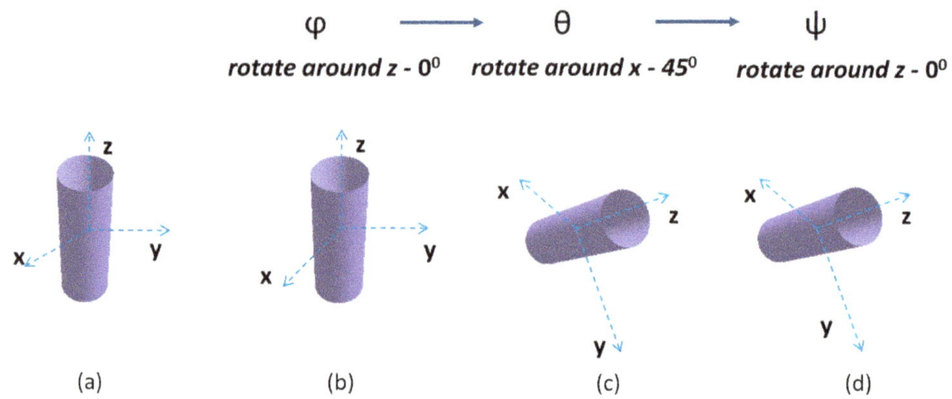

Figure 4.14. Rotation of the cylinder using the Euler angles: (a) the originally defined cylinder, (b) rotation about z at 0°, (c) rotation about x at −45°, and (d) rotation about x at 0°.

Table 4.3. Data for the four objects in the paraffin phantom.

No	Type	Property	Colour	(x_0, y_0, z_0)	Dimensions	Material	ϕ, θ, ψ
1	Parallelepiped	Phantom	Red	(0, 0, 0)	X-side = 35 mm Y-side = 55 mm Z-side = 40 mm	Paraffin	0°, 0°, 0°
2	Ellipsoid	Phantom	Blue	(0, 0, 0)	X-axis = 4.0 mm Y-axis = 4.0 mm Z-axis = 4.0 mm	Water	0°, 0°, 0°
3	Ellipsoid	Phantom	Blue	(4, −18, −4)	X-axis = 4.2 mm Y-axis = 4.2 mm Z-axis = 4.2 mm	Water	0°, 0°, 0°
4	Ellipsoid	Phantom	Blue	(8, 15, 4)	X-axis = 4.0 mm Y-axis = 4.0 mm Z-axis = 4.0 mm	Water	0°, 0°, 0°

Exercise 2: Computational phantoms based on geometrical primitives
The aim of this self-training exercise is to create a computational phantom suitable for studies with cone-beam CT. The computational phantom is composed of ten objects which are modelled as cylinders. Use click the *Save* or *Modify* button each time an object is created or modified.

Tasks: (a) create the phantom; (b) visualise it; and (c) save the phantom as *my_phantom.phn*.

Exercise 3: Computational phantoms based on geometrical primitives
The aim of this self-training exercise is to create a computational phantom dedicated to testing a dual-energy tissue cancellation algorithm. The computational phantom to be created should mimic the physical one created by Marziani *et al* (2002) and shown in figure 4.16.

Table 4.4. Data for the objects in exercise 2.

No	Type	Property	Colour	(x_0, y_0, z_0)	(Radius, height)	Material	ϕ, θ, ψ
1	Cylinder	Phantom	Blue	(0, 0, 0)	(75, 100)	Water	90°, 90°, 0°
2	Cylinder	Phantom	Blue	(0, 0, 50)	(15, 50)	Water	90°, 90°, 0°
3	Cylinder	Phantom	Red	(0, 35, 35)	(15, 50)	Cortical bone	90°, 90°, 0°
4	Cylinder	Phantom	Green	(0, 50, 0)	(15, 50)	Soft tissue	90°, 90°, 0°
5	Cylinder	Phantom	Yellow	(0, 35, −35)	(15, 50)	Adipose	90°, 90°, 0°
6	Cylinder	Phantom	Brown	(0, 0, −50)	(15, 50)	Air	90°, 90°, 0°
7	Cylinder	Phantom	Orange	(0, −35, −35)	(15, 50)	Polystyrene	90°, 90°, 0°
8	Cylinder	Phantom	Pink	(0, −50, 0)	(15, 50)	Muscle	90°, 90°, 0°
9	Cylinder	Phantom	Purple	(0, −35, 35)	(15, 50)	Gland	90°, 90°, 0°
10	Cylinder	Phantom	Gray	(0, 0, 0)	(15, 50)	Plexiglas	90°, 90°, 0°

Figure 4.15. Solid-based phantom for use in dual-energy studies.

This phantom consists of a polymethyl methacrylate (PMMA) slab which includes partially superimposed cylinders filled with either polyethylene (PE) or water. The dimensions are depicted in figure 4.16. The materials and the relative positions of the inserts were chosen in order to test the ability of the above algorithm to cancel contrasts arising from materials with similar x-ray transmission properties (Marziani *et al* 2002).

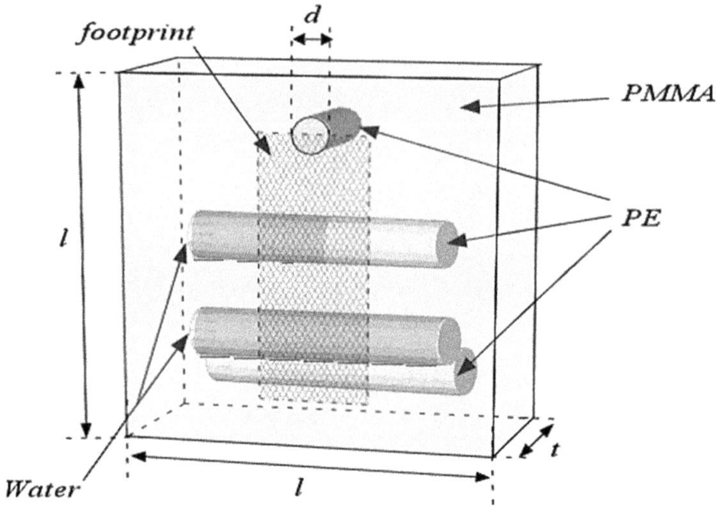

Figure 4.16. Schematic representation of the plastic phantom and the details included. The dimensions are: l = 50 mm, t = 20 mm, and d = 6 mm. All cylinders have the same diameter d. The size of the irradiated area (footprint) was 6.6 mm × 23.4 mm. (Reproduced from Marziani *et al* 2002 with the permission of the publisher.)

Table 4.5. Parameters of the Shepp–Logan phantom (Shepp and Logan 1974).

No	a	b	c	x	y	z	θ	ϕ	Absorption coefficient
1	0.690	0.9	0.92	0.0	0.0	0.0	0.0	0.0	2.0
2	0.662	0.88	0.874	0.0	0.0	−0.0184	0.0	0.0	−0.98
3	0.410	0.21	0.16	−0.22	−0.25	0.0	0.0	72.0	−0.02
4	0.310	0.22	0.11	0.22	−0.25	0.0	0.0	−72.0	−0.02
5	0.210	0.35	0.25	0.0	−0.25	0.35	0.0	0.0	0.01
6	0.046	0.046	0.046	0.0	−0.25	0.1	0.0	0.0	0.01
7	0.046	0.02	0.023	−0.008	−0.25	−0.605	0.0	0.0	0.01
8	0.046	0.02	0.023	0.06	−0.25	−0.605	0.0	90.0	0.01
9	0.056	0.1	0.04	0.06	0.625	−0.105	0.0	0.0	0.02
10	0.056	0.1	0.056	0.0	0.0625	0.1	0.0	0.0	−0.02
11	0.046	0.046	0.046	0.0	−0.25	−0.1	0.0	0.0	0.01
12	0.023	0.023	0.023	0.0	−0.25	−0.605	0.0	0.0	0.01

Tasks: (a) create the phantom; (b) visualise it; and (c) save the phantom in *phn* format.

Exercise 4: Computational phantoms based on geometrical primitives.
The aim of this self-training exercise is to design a Shepp–Logan phantom (Shepp and Logan 1974), which serves as a model of a human head in the development and testing of image reconstruction algorithms. The phantom consists of 12 ellipsoids with semi-axes as specified in table 4.5. Their position, rotation, and density are also specified in the same table.

Figure 4.17. Shepp–Logan phantom, implemented with the objects described in table 4.5.

The implementation of the objects described in table 4.5 will result in a phantom as shown in figure 4.17.

Tasks: (a) create the phantom; (b) visualise it; and (c) save the phantom in *phn* format.

Exercise 5: A computational CIRS 011A phantom based on geometrical primitives.
The aim of this self-training exercise is to design a computational replica of the CIRS phantom, shown in figure 2.11(b), chapter 2, which is a model of compressed breast and is used as a QA phantom. The detailed description available at http://www.cirsinc.com/products/solution/6/tissue-equivalent-phantom-for-mammography/ is summarised in table 4.6.

Tasks: (a) create the phantom; (b) visualise it; and (c) save the phantom in *phn* format.

4.4 Creating anthropomorphic phantoms via *LUCMFRGen*

In the research there is a need for phantoms that are composed of identical solid objects, the number of which may be as large as several thousands. One such example is the physical phantom developed at the Department of Radiology, Catholic University of Leuven, that produces a structured background in both two-dimensional mammography and breast tomosynthesis (shown in figure 2.16(a), chapter 2). This physical breast phantom is based on the work of Gang *et al* (2010),

Table 4.6. Description of details inserted in CIRS 011A.

Object	No	Description
Line pair target	1	5–20 lp/mm: 17 segments with 5 lines each
$CaCO_3$ specs grain size (mm)	2	0.130
	3	0.165
	4	0.196
	5	0.230
	6	0.275
	7	0.400
	8	0.230
	9	0.196
	10	0.165
	11	0.230
	12	0.196
	13	0.165
Step wedge 1 cm thick	14	100% gland
	15	70% gland
	16	50% gland
	17	30% gland
	18	100% adipose
Nylon fibre diameter size (mm)	19	1.25
	20	0.83
	21	0.71
	22	0.53
	23	0.30
Hemispheric masses 75% glandular/25% adipose thickness (mm) (model 010B, 010D, 011A), 55% glandular/45% adipose thickness (mm) (model 010A, 010C)	24	4.76
	25	3.16
	26	2.38
	27	1.98
	28	1.59
	29	1.19
	30	0.90
Optical density	31	Reference zone
Edge of beam	32	Localisation target
Phantom specifications		Dimensions: 2.5 cm × 18.5 cm; weight: 2 lbs (0.7 kg)
Materials		Phantom: epoxy resin; specs: calcium carbonate
Models 010 and 011A include		Tissue-equivalent mammography phantom adipose tissue-equivalent layer (0.5 cm thick)

who showed that equal volumes of differently sized acrylic spheres provide a fractal dimension of 3 as well as a power law exponent β equal to 3. The phantom of Gang *et al* is shown in figure 4.18.

Based on this phantom, the physical LUCMFR phantom was created, composed of two main parts: an acrylic semi-cylinder container of thickness 58 mm and diameter of 200 mm, and equal volumes of acrylic spheres of six different diameters: 15.88 mm, 12.70 mm, 9.52 mm, 6.35 mm, 3.18 mm, and 1.58 mm (Cockmartin *et al* 2012). The spherical objects are placed within the semi-circular container which resembles a compressed breast. As an alternative to spheres placed in air, the space between the spherical objects can also be filled with water, a material with x-ray properties close to those of real breast tissue. This phantom is shown in figure 2.16(a), chapter 2. This physical phantom is characterised by air bubbles which were visible at the upper side of the phantom because of the practical difficulties in completely filling the phantom with water. To further improve this phantom through virtual clinical trials, a software application, *LUCMFRGen* has been created, which can generate containers of different shapes: slab, cylindrical, semi-cylindrical, or spherical, filled with a texture created from semi-ellipsoids or ellipsoids with different dimensions and tissue-equivalent material.

4.4.1 *LUCMFRGen* software application

The *LUCMFRGen* application is an application that generates different shaped containers—cylindrical, spherical, semi-cylindrical, and slabs filled with spheres or ellipsoids of different dimensions and elemental compositions.

Figure 4.18. Phantom used in the study of Gang *et al* (2010). (Reproduced from Gang *et al* 2010 with the permission of John Wiley & Sons. © 2010 American Association of Physicists in Medicine.)

The application file for Microsoft Windows OS, with relevant help and demo files, will be available for the participants in the EUTEMPE course.

An example of the semi-cylindrical container is demonstrated below. The container may be modelled as a set of four semi-cylinders and a slab object (approximating the sidewall) as shown in figure 4.19(a). The top and the base of the phantom, however, may be included in the external semi-cylinder, which reduces the container model to two semi-cylinders and one slab object, as shown in figure 4.19(b).

The sphere generation is accomplished as follows. Initially, the position of the spherical objects within the internal semi-cylinder is sampled randomly, as this procedure starts from the spheres with the largest diameters. Then, the developed algorithm verifies that the sampled sphere does not intersect with other spheres at the sampled position. A screenshot of this tool is shown in figure 4.20(a). It is distinguished by two main fields: one of the fields is dedicated to the design of the container dimensions (wall thickness, diameter, and height) and material characteristics, while the other field is dedicated to the spherical objects. For example, the phantom set-up shown in figure 4.20(a) corresponds to the creation of a software breast phantom with the same parameters as the physical one shown in figure 2.16(a), chapter 2. The statistics, shown at the bottom of the tool, indicate that the total number of generated spherical objects is 30 272, and 26 of these are spheres with a diameter of 15.88 mm. The created phantom is shown in figure 4.20(b).

A major advantage of this software application is the possibility to create phantoms with different dimensions and content, as well as the possibility to assign various material characteristics to the simulated objects. The number of different spheres is six. The software tool may be used to study the appearance of simulated planar images of generated breast phantoms with different characteristics.

Another case is the use of other container shapes, such as the spherical container, shown in figure 4.21.

4.4.2 Self-training exercises

Exercise 1: Computational phantoms based on geometrical primitives

The aim of this self-training exercise is to design a three-component breast phantom dedicated to dual-energy analysis (Taibi *et al* 2003). The physical phantom is shown

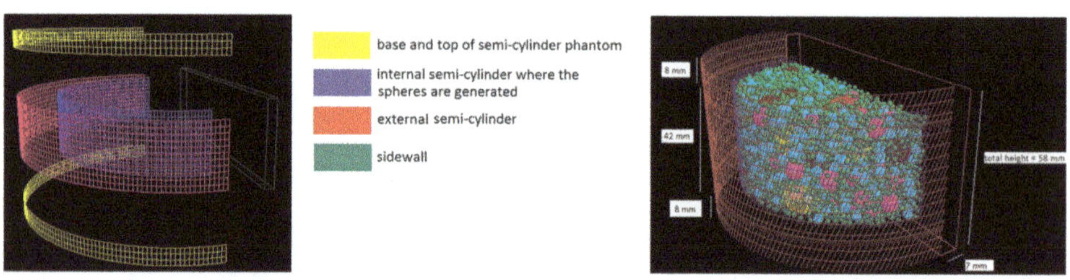

(a) (b)

Figure 4.19. The software breast phantom: (a) external shape components and (b) the complete breast phantom (Bliznakova 2016).

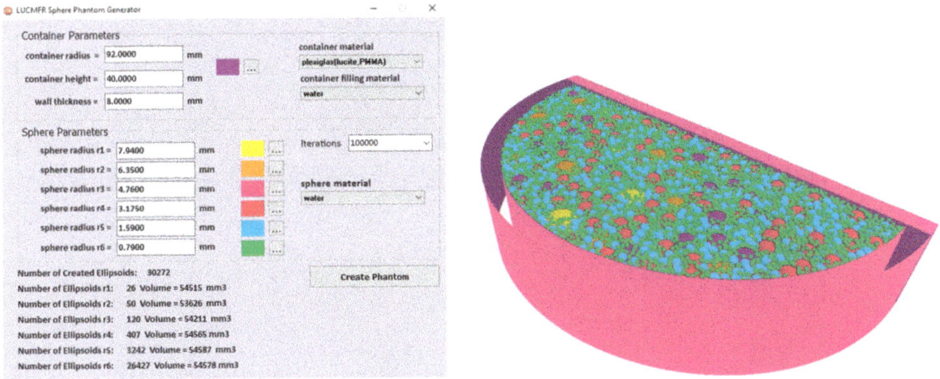

(a) (b)

Figure 4.20. (a) A screenshot of the software application *LUCMFRGen* showing the design of a software breast phantom with similar characteristics to the physical prototype and (b) the created software phantom, opened and visualised with *XRAYImagingSimulator*.

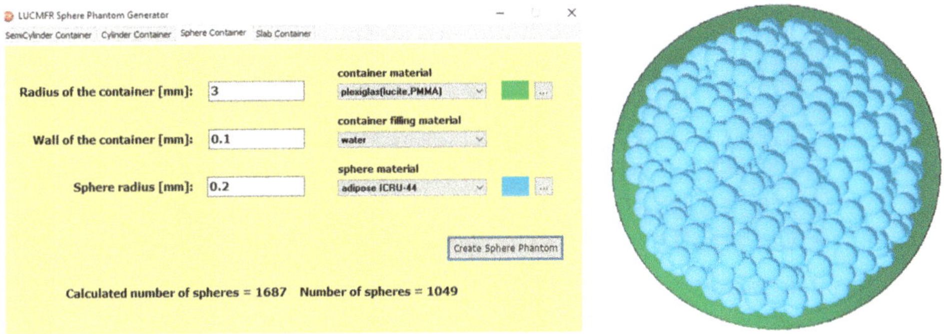

(a) (b)

Figure 4.21. Creating a spherical container filled with spherical objects.

in figure 4.22. The phantom consists of a Plexiglas box 5 cm thick, filled with water and Plexiglas spheres of three different diameters, 8 mm, 12.7 mm, and 22.2 mm. Five polyethylene spheres are embedded in the phantom as details, with diameters ranging from 3.2 mm to 12.7 mm. The bigger polyethylene details are clearly visible as white spheres.

Tasks: (a) create the phantom by using the *LUCMFRGen* tool; (b) visualise it; and (c) save the phantom in *phn* format.

Exercise 2: Computational phantoms based on geometrical primitives
The aim of this self-training exercise is to generate a phantom for testing tomosynthesis imaging. The phantom, shown in figure 4.23, represents a container of paraffin, filled with 155 water spheres. The diameter of the spheres is 8 mm, while the size of the container is 60 mm × 60 mm × 40 mm (width, length, height).

Figure 4.22. A three-component breast phantom dedicated to dual-energy studies (Reproduced from Taibi *et al* 2003 with the permission of the publisher).

Tasks: (a) create the phantom; (b) visualise it; and (c) save the phantom in *phn* format.

Exercise 3: Computational phantoms based on geometrical primitives
The aim of this self-training exercise is to design a phantom similar to the original LUCMFR phantom (Cockmartin *et al* 2012) using the *LUCMFRGen* tool.

Tasks: (a) create a small sized container, e.g. 10 × 10 × 1 mm, and four different radii of spheres: 4 to 1 mm. The material for the container can be paraffin and for the spheres water; (b) visualise the phantom using the *XRAYImagingSimulator*. The created phantom should be similar to the one shown in figure 4.24.

4.5 Creating voxel-based anthropomorphic phantoms

Two approaches for the creation of voxel phantoms are discussed: (a) based on 3D medical images of patients and physical phantoms and (b) by transforming the CSG to voxel geometry. Both are implemented using the *XRAYImagingSimulator*. Self-study

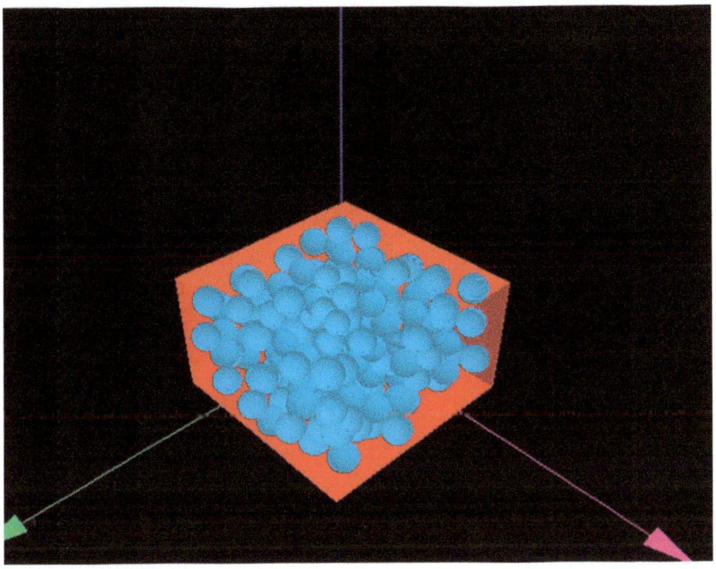

Figure 4.23. A screenshot of the 3D visualisation of the paraffin–water phantom. Size of the container: [*x*, *y*, *z*] 60 mm × 60 mm × 40 mm. 155 water spheres are inside the container.

video tutorials are available at the following link: http://tuv.eutempe-net.eu/wp-content/uploads/Educationmaterials/Videos/Createvoxelisedphantom-1%20optimized.mp4.

4.5.1 Tutorial on the creation of a voxel-based anthropomorphic phantom from medical images

Medical images from 3D imaging techniques are suitable for the creation of voxel-based phantoms. The following tutorial explains step by step how to create a 3D voxel-based phantom from patient CT abdomen images. In general, medical images are stored in DICOM format. The format of the image is unsigned short integer values [0 to $2^{16}-1$]. Sometimes, physicians deliver the patient data in other imaging formats such as *bmp*, *png*, etc. The main idea is to perform a tissue classification algorithm, which can be written in scripts (*MATLAB*, *IDL*, *ImageJ*, etc) or be implemented in a user-friendly application.

To create a voxel phantom with the *XRAYImagingSimulator*, the option *Create Voxelised Phantom* from the *Phantom* menu is used (figure 4.4). The user selects the folder with the medical images (figure 4.25). Selecting each one of these images results in their visualisation. The important information is the size of the obtained voxel phantom. This information is available in the lower part of the screen shown in figure 4.25. The size of the voxel phantom is described by its width, length, and height. The voxel phantom is built by using the *Build Phantom* button. Then, a tissue classification algorithm is applied on these images, which results in a voxel phantom with clearly differentiated tissues.

Figure 4.24. A phantom generated with the help of the *LUCMFRGen* tool.

In the particular case shown in figure 4.25, the images are the slices from patient CT of the abdominal area. These images are of size 512 × 512 pixels and 8 bit grey level resolution. Some slices are shown in figure 4.26. The pixel resolution is 1 mm. These data are in the folder *Phantoms/CT slices/VoxelLiver1*.

After image slices are stacked in a phantom, the segmentation of tissues is implemented. For this purpose, *Option* → *Tissue values* (figure 4.27(a)) is used to segment air, water, soft tissues, and bone (for this particular case).

The screen shots in figure 4.28 reflect the procedure of segmentation of tissues in CT liver images. In short, this procedure may be described with the following pseudo-code:

if (*pixel value* == 0) *tissue* = *air*
if (*pixel value* > 0 && *pixel value* <= 60) *tissue* = *water*
if (*pixel value* > 60 && *pixel value* <= 100) *tissue* = *soft tissue*
if (*pixel value* > 100 && *pixel value* <= 255) *tissue* = *bone*

where *pixel value* is the value of the pixel in the original CT image.

When the segmentation is performed, the voxel phantom must be saved using the *Save* option (figure 4.27(b)). For instance, save the phantom as *VoxelLiver1.bin.* The format of the data is 16 bit unsigned, while the file extension is **.bin*.

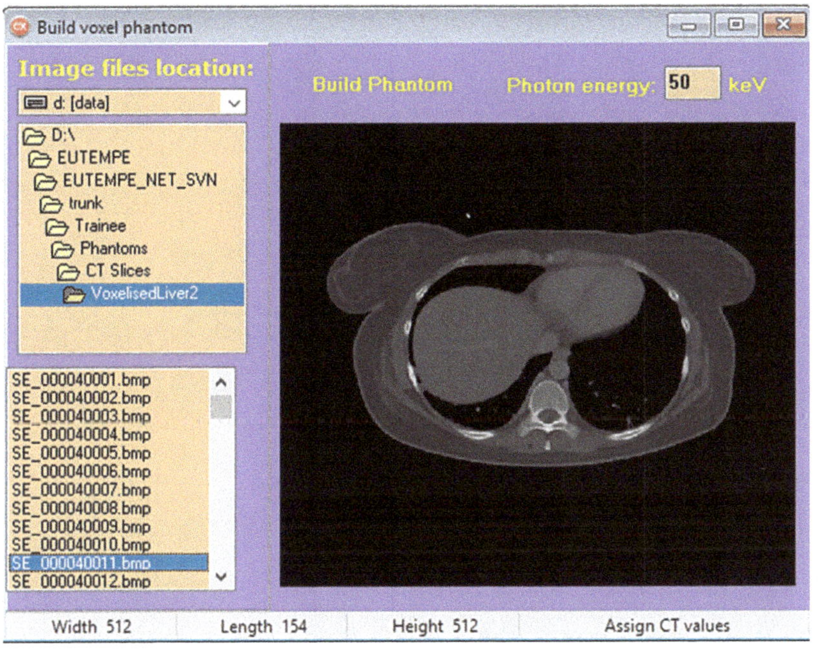

Figure 4.25. Module used for building a voxel phantom.

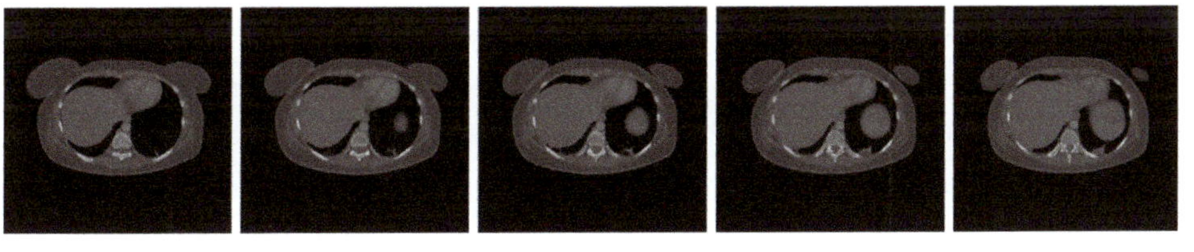

Figure 4.26. Consecutive slices from patient abdominal CT.

Figure 4.27. Options in the module for the creation of a voxel phantom: (a) segmentation of tissues in patient tissues and (b) saving the voxel phantom.

Data are stored in raw format, without a header. For instance, using the *ImageJ* application, the created voxel phantom is opened by using the settings shown in figure 4.29(a). Figure 4.29(b) shows the phantom using one of the plug-ins for 3D image visualisation.

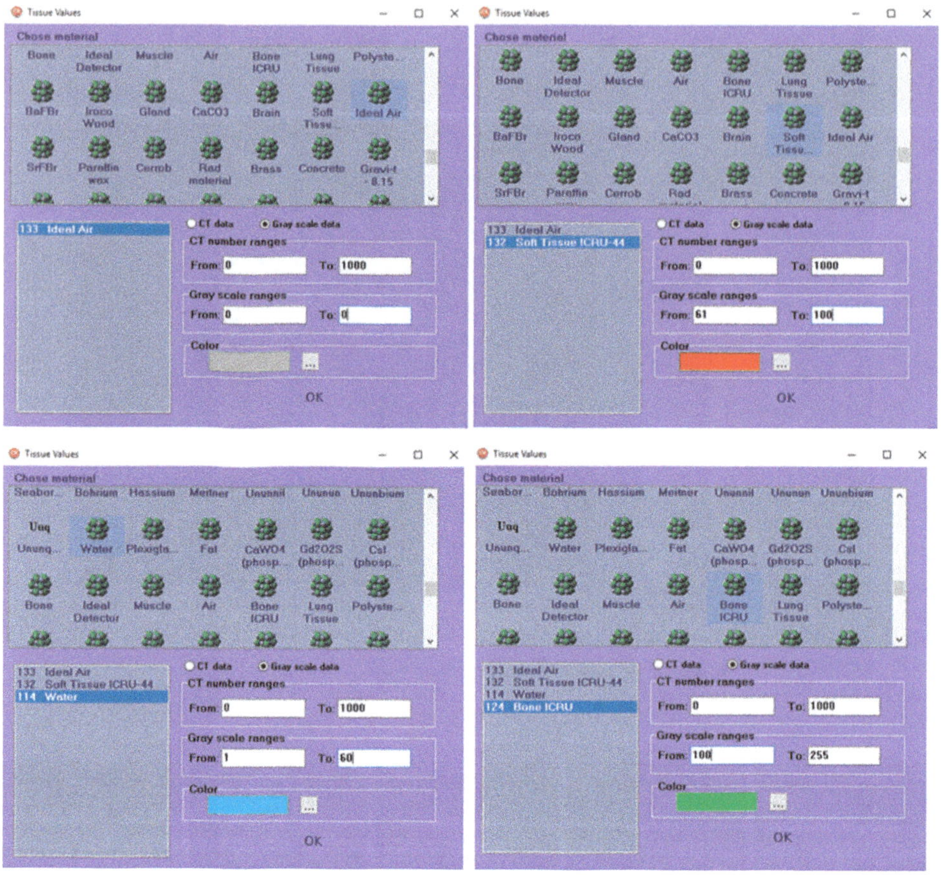

Figure 4.28. Example screens from *Tissue Values* used to segment tissues from CT images of the patient liver.

4.5.2 Tutorial on the creation of a voxel-based anthropomorphic phantom from geometrical primitives

Voxel phantoms may be created from phantoms built from geometrical primitives. The main steps for the creation of such a phantom are demonstrated for a simple solid object below:

(a) Create a sphere with a radius of 3 mm using the *New Phantom* module. Assign water composition to this sphere.
(b) Save the sphere as *small_sphere.phn.*
(c) Use the *Voxel based phantom* option from this module to voxelise the created phantom (figure 4.30a), first row).
(d) The invoked panel is shown in figure 4.30(a), second row. The voxels in the voxelised phantom will have dimensions specified in the *Voxel Size* width, length, and height combo boxes, while the size of the 3D matrix in pixels is shown in the *Calculated Voxel Matrix* width, length, and height edit boxes.
(e) Voxelise this sphere with a voxel size of 1 mm.

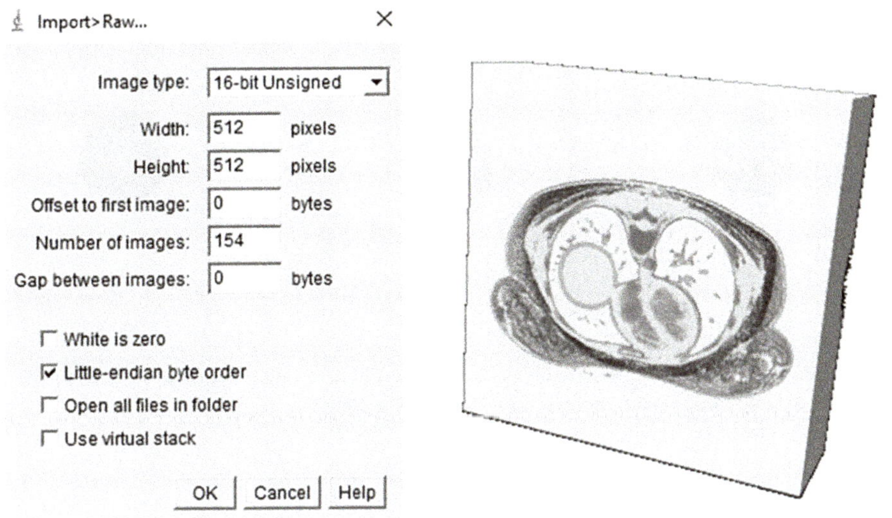

(a) (b)

Figure 4.29. Use of *ImageJ* to open the built voxel phantom: (a) settings and (b) voxel phantom visualisation.

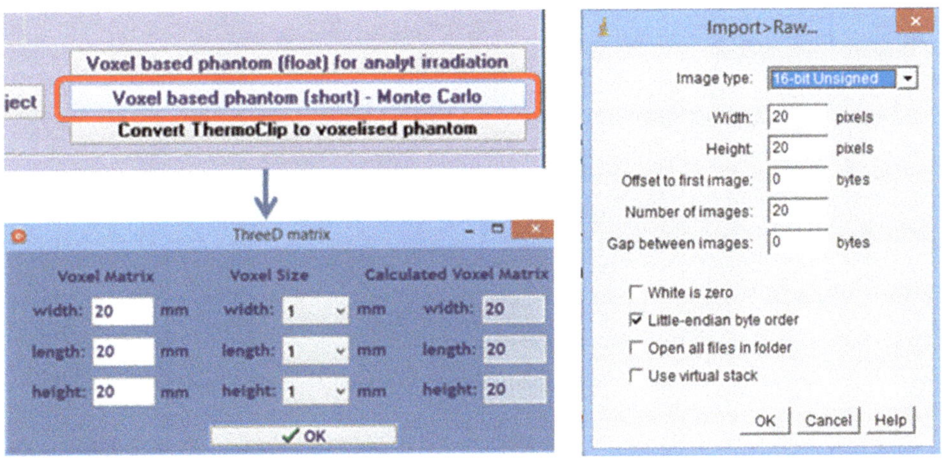

(a) (b)

Figure 4.30. Voxelisation of CSG phantoms: (a) voxelisation and (b) opening the voxelised phantom by *ImageJ*.

(f) The voxel phantom is stored in the */Phantoms* directory and has the same name as the small_sphere.phn file with the *.bin* extension (*small_sphere.bin*).

(g) To visualise the voxel phantom (*small_phantom.bin*) the *ImageJ* application is used. Specifically, *ImageJ* → *Import* → *Raw* with the parameters shown in figure 4.30(b). The result is shown in figure 4.31(a). The data format is unsigned short.

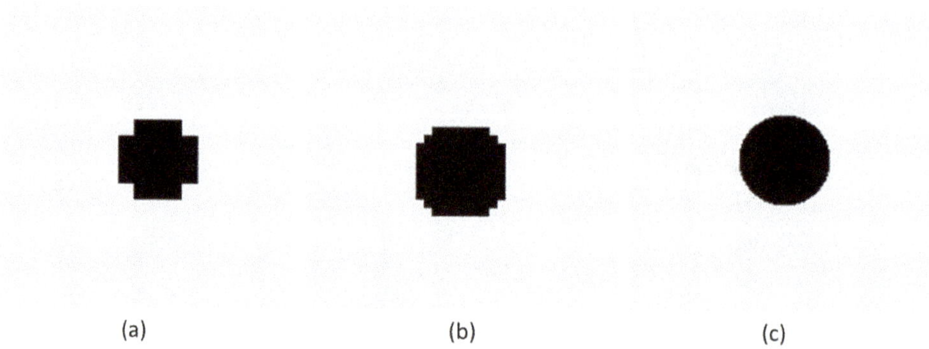

Figure 4.31. Slices from small_sphere.bin phantom: voxel size of (a) 1 mm (volume 20 × 20 × 20), (b) 0.5 mm (volume 40 × 40 × 40), and (c) 0.1 mm (volume 200 × 200 × 200).

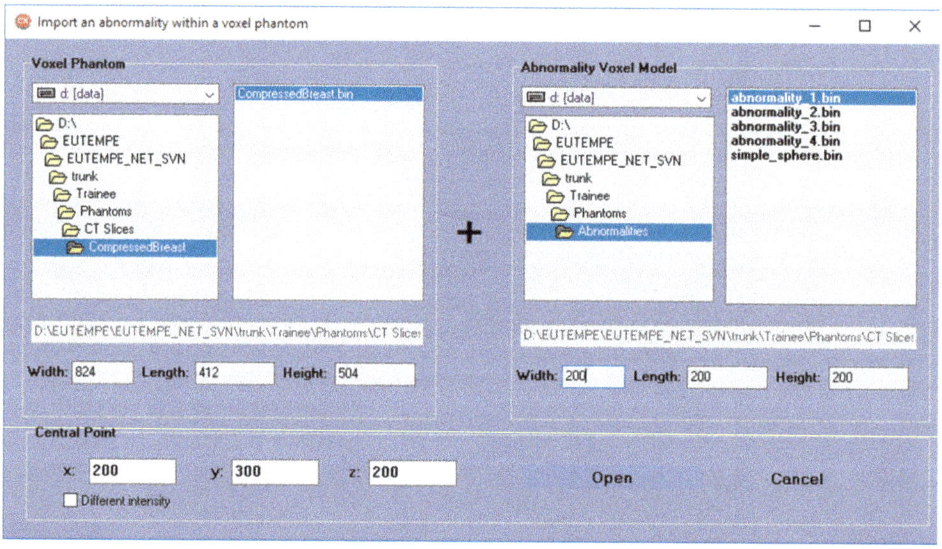

Figure 4.32. A screenshot from the module used to insert voxel models of abnormalities into another voxel phantom.

(h) Repeat the same experiment for voxel sizes 0.5 mm and 0.1 mm. In these two cases, the matrix size is 40 × 40 × 40 pixels and 200 × 200 × 200 pixels, respectively. Slices from the obtained volumes are shown in figures 4.31(b), (c).

4.5.3 Tutorial on inserting created lesions into a voxel-based anthropomorphic phantom

In many virtual clinical studies, a voxel model of an abnormality is required to be inserted into the computational model of healthy tissue volume. For this purpose, the *Tools→Add an abnormality* option is used. An example is shown in figure 4.32 for the case of inserting a voxel model of an abnormality within a voxel breast phantom. In this specific case, the breast phantom has dimensions of 824 (width) × 412 (length) × 504 (height). The abnormality has dimensions of 40 (width) × 65 (length) × 106 (height) and it is inserted in the point defined by the (x, y, z) values specified in the edit boxes of the *Central Point*. The result of this procedure is shown in figure 4.33.

4.5.4 Self-training exercises

Exercise 1: Creation of a voxel phantom from patient-specific breast data

The aim of this self-training exercise is to create a voxel-based software breast model with a realistic breast tissue distribution based on segmented patient CT breast images. For this purpose, slices are obtained from a dedicated breast CT scanner at the Department of Radiology and Imaging Sciences at Emory University (Atlanta, GA, USA). These data are de-noised and segmented in order to obtain a compositional breast model composed of skin, glandular, and adipose tissues (Yang *et al* 2011). An example of a processed slice is given in figure 4.34(a), while values in slices are listed in the table shown in figure 4.34(b). The voxel size of the 3D breast model is 0.28 mm in each direction.

The data are in folder */Phantoms/CT slices/VoxelBreast1*.

Tasks:

(a) Use the *XRAYImagingSimulator* and create the voxelised phantom model. From the main menu select, *Phantom* → *Create Voxel Phantom*.

(b) Save the phantom as *VoxelBreast1.bin.* Data are saved in 16 bit unsigned format.

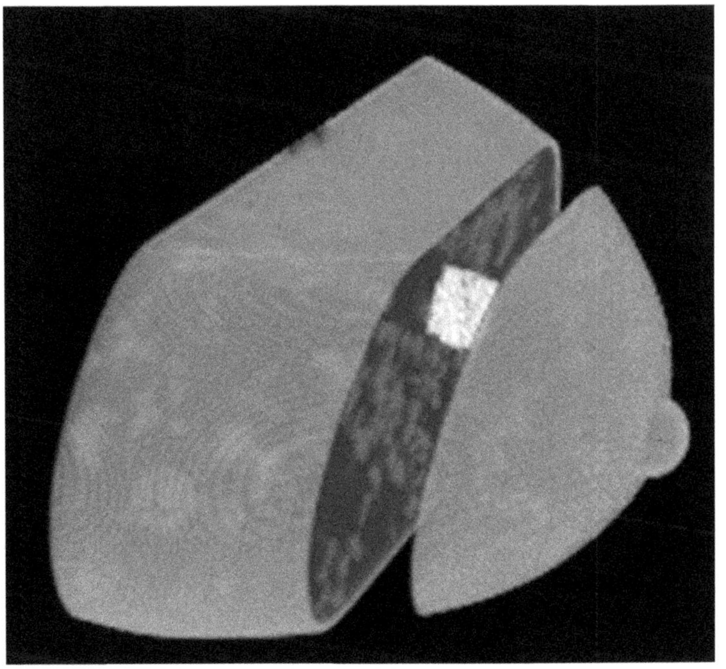

Figure 4.33. Breast voxel phantom with an imported abnormality using the interface shown in figure 4.32. (Reprinted from Dukov *et al* 2019 with the permission of the publisher).

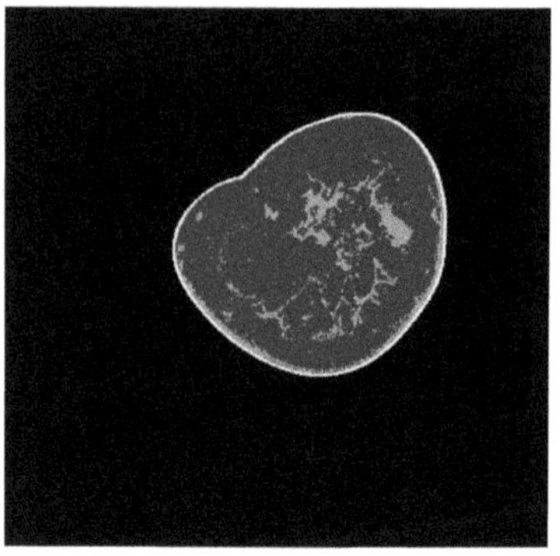

(a)

tissue	value
Skin	255
Air	0
Gland	170
Adipose	85

(b)

Figure 4.34. (a) Processed slice from a breast CT acquisition set and (b) values in the segmented CT slice.

(c) Visualise the phantom with the help of *ImageJ*. The format of the matrix is 16 bit unsigned.

Exercise 2: Creation of a voxel phantom from patient-specific abdomen data
The aim of this self-training exercise is to create a voxel phantom based on patient CT abdomen images. The data for this exercise are in folder */Phantoms/CT slices/VoxelLiver2*.

Tasks:

(a) Use the *XRAYImagingSimulator* and create the voxelised phantom model.
(b) For this purpose use *Option* → *Tissue values* to set the values for air, soft tissues, and bone.
(c) Save the phantom as *VoxelLiver2.bin.*
(d) Visualise the phantom with the help of *ImageJ*.

Exercise 3: Creation of a voxel phantom from a phantom based on geometrical primitives
The aim of this self-training exercise is to create a voxel-based phantom from a phantom built from geometrical primitives. Create a CSG phantom as a combination of two 3D shapes: parallelepiped and ellipsoid. Example parameters are the following:

Object 1 parallelepiped:	Centre point 0, 0, 0	Dimensions: 40 mm, 40 mm, 40 mm
	φ, θ, ψ: 0°, 0°, 0°,	Composition: adipose tissue
Object 2 ellipsoid:	Centre point 0, 0, 0	Dimensions: 4 mm, 5 mm, 6 mm
φ, θ, ψ: 0°, 0°, 0°,	Composition: glandular tissue	

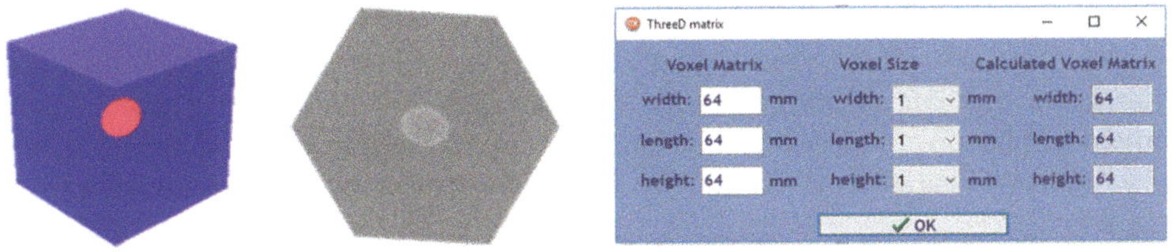

(a) (b) (c)

Figure 4.35. Creation of a voxel phantom based on a phantom built from geometrical primitives: (a) the original phantom built from a cube with a size of 40 mm × 40 mm × 40 mm and an ellipsoid with the following semi-axis: 4 mm × 5 mm × 6 mm. (b) The voxel phantom obtained using the options in (c).

With these parameters the phantom shown in figure 4.35(a) is obtained.

Its voxel version is obtained by using the *Voxel based phantom (short)—Monte Carlo* option. The voxel version shown in figure 4.35(b) is obtained by using the options shown in figure 4.35(c).

Tasks:

(a) Use the *XRAYImagingSimulator* and create the phantom from geometrical primitives.
(b) Save the phantom as *simple.phn*.
(c) Use the option *Voxel based phantom (short)—Monte Carlo* to convert this phantom to a voxel-based phantom.
(d) Use the options shown in figure 4.35(c).
(e) Visualise the *simple.bin* phantom with the help of *ImageJ*.

4.6 Simulation of x-ray image formation

4.6.1 Basics of x-ray image formation

4.6.1.1 Attenuation of x-rays

When an x-ray photon interacts with an atom in the absorption process, part of its energy E_{ph} is converted into the kinetic energy of high-speed charged particles (electrons or positrons), while part of the energy is radiated from the absorber as scattered radiation. The energy deposited per unit mass of the medium is known as the absorbed dose and is a very useful quantity for the prediction of biological effects (Johns and Cunningham 1983). The scattered x-rays possess new energy and directions of propagation, depending on the interaction type. X-ray photons can be subjected to any of the four x-ray interaction types: (i) coherent scattering; (ii) Compton scattering; (iii) photoelectric absorption; and (iv) the pair production process. High-energy electrons and positrons are produced by the last three photon interactions and constitute charge transport, which is essential to the energy redistribution and deposition in the absorber. The charged particles, while passing through matter, undergo a large number (10^6) of elastic and inelastic collisions with the nuclei and the atomic electrons until their final fate. These include Møller and bremsstrahlung interactions for the electrons and Bhabha and annihilation interactions for the positrons.

In this chapter, we will focus on the transport of x-rays with energies up to 100 kVp. Figure 4.36(a) shows a diagram, illustrating how a material of thickness x reduces the number of photons reaching the detector placed at a point P. The removal of photons from a beam of x-rays as it passes through the material is called *attenuation of x-rays*. It is caused by both absorption and scattering of primary photons.

Let the number of photons recorded at point P be N (figure 4.36(b)), while the initial number of photons is N_0. The number of photons, n, removed from the beam due to the radiation interaction with the matter is:

$$n = \mu N_0 x, \tag{4.3}$$

where μ is the *linear attenuation coefficient*, typically expressed in cm^{-1}. The attenuation coefficient μ is expressed from equation (4.3) as follows:

$$\mu = \frac{n/N_0}{x} = \frac{\text{the fraction of photons that interact in } x}{x}. \tag{4.4}$$

The linear attenuation coefficient is also a sum of individual linear attenuation coefficients for each type of interaction:

$$\mu = \mu_{\text{Coherent}} + \mu_{\text{photoelectric}} + \mu_{\text{Compton}} + \mu_{\text{pair\_production}}. \tag{4.5}$$

For a monoenergetic beam of photons incident on either thick or thin slabs of material, an exponential relationship exists between the number of incident photons, N_0 and those transmitted, N, through thickness x without having interaction with the material (figure 4.36(b)):

$$N = N_0 e^{-\mu x}, \tag{4.6}$$

which is the so-called *Beer–Lambert law*. Simulation of image formation is implemented in the *XRAYImagingSimulator* using this law.

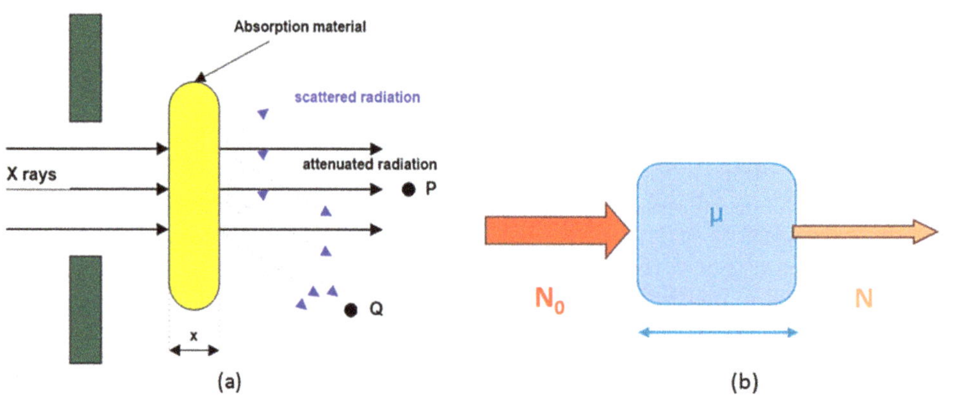

Figure 4.36. Attenuation of x-rays: (a) a diagram illustrating how an attenuator of thickness x reduces the number of photons reaching P which is a point on the detector and (b) a simplified scheme of the attenuation of N_0 initial number of x-rays when passing through a material with a linear attenuation coefficient μ.

4.6.1.2 Modelling the image acquisition

This includes the acquisition scheme and the detector. The detector may be modelled as a parallelepiped with specific dimensions, as well as number of pixels and pixel resolution, and the physical properties of the detector: atomic number, mass, density, and interaction coefficients. Figure 4.37 shows a model of a detector.

For the purposes of the EUTEMPE course, the use of the detector model is omitted. This model is used with Monte Carlo simulations. The obtained image, with characteristics specified in figure 4.41, is a noise-free image. Additional noise might then be added to it in order to achieve images more realistic than those obtained in clinical practice.

4.6.1.3 Ray tracing

Once the x ray system and the patient are designed, the acquisition scheme, i.e. the source–detector and object–detector distances, the acquisition arc, and the number of projection images, need to be specified. Figure 4.38(a) depicts a scenario for a virtual study, with a patient represented as an elliptical cylinder lying on the patient table, and undergoing imaging from two positions of the x-ray tube. The tube and the detector rotate synchronously about the patient. From this sketch it is seen that the main parameters needed to uniquely model the acquisition are: x-ray spectra, image resolution, acquisition arc, distances from the source to the patient and the detector, patient model, etc.

The simulation of x-ray beam transport from the x-ray tube through the patient reaching the detector, results in the generation of 2D images. This simulation exploits the concept of ray tracing, as this includes the definition of the ray and its intersections with the phantoms. An example is shown in figure 4.38(b). The source emits rays with intensity I_0 towards the detector. Using the ray tracing concept, the intersection points of the ray with each object along its path are determined. The path lengths through each of the regions that the ray travels in the objects (t_1, t_2, t_3) are calculated. The transmitted intensity reaching the detector pixel is calculated using the Beer–Lambert law (equation (4.6)) but adapted for intensity images:

$$I = I_0 * \exp\left(-\int_{ll} \mu(x,\, y,\, z)dl\right), \tag{4.7}$$

where $\mu(x, y, z)$ is the spatially dependent linear attenuation coefficient, l is the path length through the object, and I_0 is the intensity of radiation at the source segment

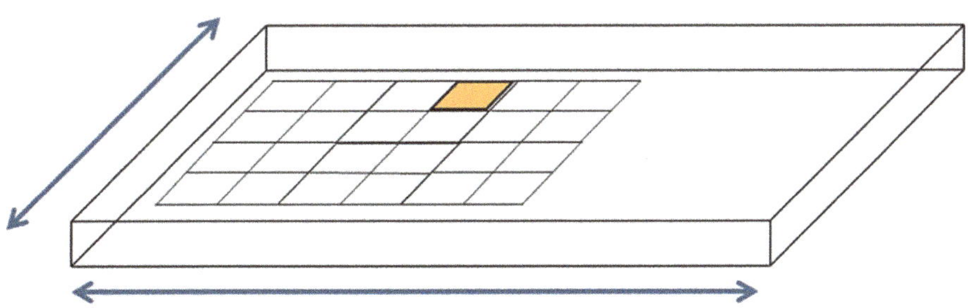

Figure 4.37. Detector modelling approach.

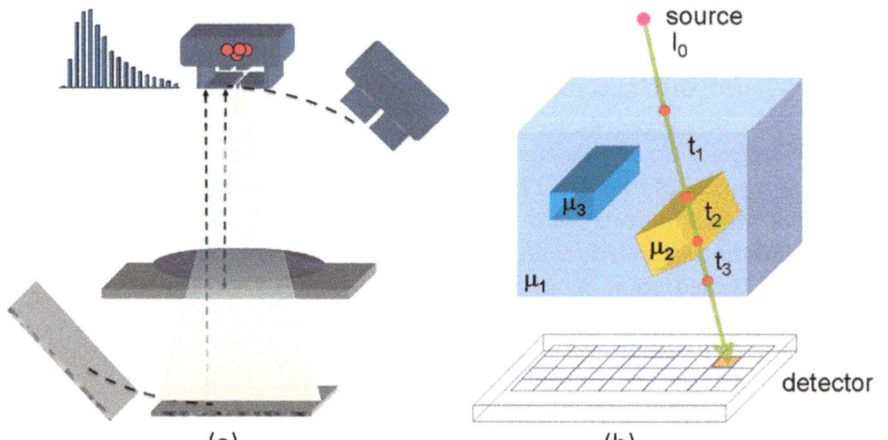

Figure 4.38. Modelling and simulation of x-ray imaging: (a) a sketch demonstrating x-ray imaging techniques and (b) ray tracing used to simulate the radiation transport.

that emits to the area of the detector, taken as equal for all viewing angles. Images obtained with this approach are called analytical or free of scatter, since they exploit the analytical relationships of x-ray–matter interaction.

Figure 4.39 depicts three examples of acquisition geometries that can be easily modelled and simulated: (a) isocentric acquisition geometry, (b) partial isocentric motion with a source rotating around the patient and a detector which moves in a plane, and (c) partial isocentric motion with a still detector and source rotating around the patient.

A more complex level of simulation of the radiation interaction at the absorber and the detector includes Monte Carlo techniques. These are used to calculate the photon transport by sampling interactions with the matter and the distances that x-rays travel until the next interaction. Since these techniques are time-consuming, the use of powerful multi-core computers, GRID, or CLOUD infrastructure is necessary to provide timely results.

Figure 4.40 shows the two options from the *Analytical Image Simulation* menu for simulation of image formation dedicated to solid-based geometry and voxel phantoms.

The following sections focus on the use of these two options with phantoms built from geometrical primitives and voxel phantoms.

4.6.2 Tutorial for image formation from solid-based anthropomorphic phantoms

The module is used to model the acquisition geometry by specifying distances from the source to the isocenter and detector, gantry angles, image size, pixel resolution, etc. Acquisition of mammography, tomosynthesis, CT, helical CT images, and radiographic images of the computational phantoms may be modelled and simulated. The x-ray beam may be parallel or fan. X-ray projection images are obtained with simulation of x-ray photon transport starting from the x-ray source, passing

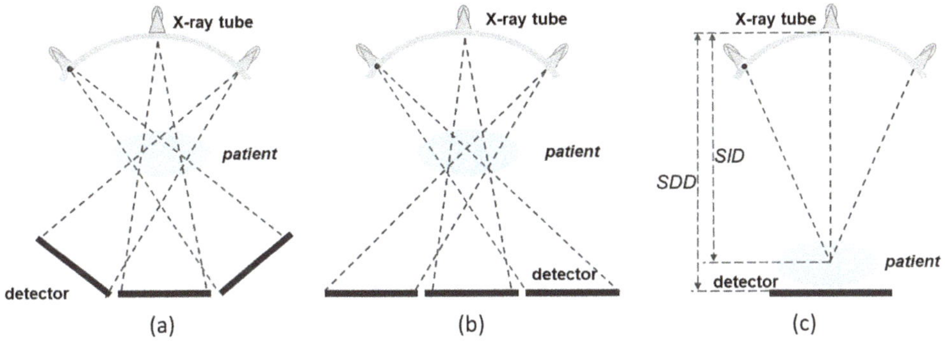

Figure 4.39. Example of simulated geometries: (a) isocentric acquisition geometry, (b) partial isocentric motion with a source rotating around the patient and detector, which moves in a plane, and (c) partial isocentric motion with a stationary detector and source rotating around the patient.

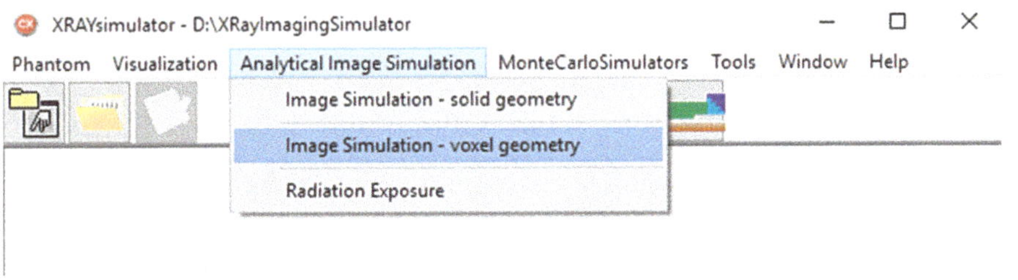

Figure 4.40. The *Analytical Image Simulation* menu used to model and simulate image formation through CSG and voxel computational phantoms.

through the breast model and reaching the detector. The pixel values are line integrals expressed as follows:

$$\sum_{i=1}^{n} \mu_i d_i, \tag{4.8}$$

where the parameters μ_i and d_i are the attenuation coefficient and the path length through each object out of a total of n objects from the phantom encountered.

Before using the *Image Simulation—solid geometry* option (figure 4.39), which is dedicated to x-ray imaging simulations of solid-based geometry objects, it is necessary to open an existing solid-based phantom or create a new one. A self-study video tutorial is available at the following link: http://tuv.eutempe-net.eu/wp-content/uploads/Education-materials/Videos/SolidGeometryImageFormation.mp4.

A screenshot of the module for image simulation of CSG is shown in figure 4.41. This set-up corresponds to a limited-arc image acquisition with 21 projection images registered every two degrees in the tube angular position range from −20° to 20°, a detector placed 130 cm away from the x-ray source, and a monoenergetic fan beam with an energy of 20 keV. The data are stored in folder */DataA* in double precision. Their size is 256 × 256 pixels and may be viewed in *ImageJ* by using the settings shown in figure 4.42(b).

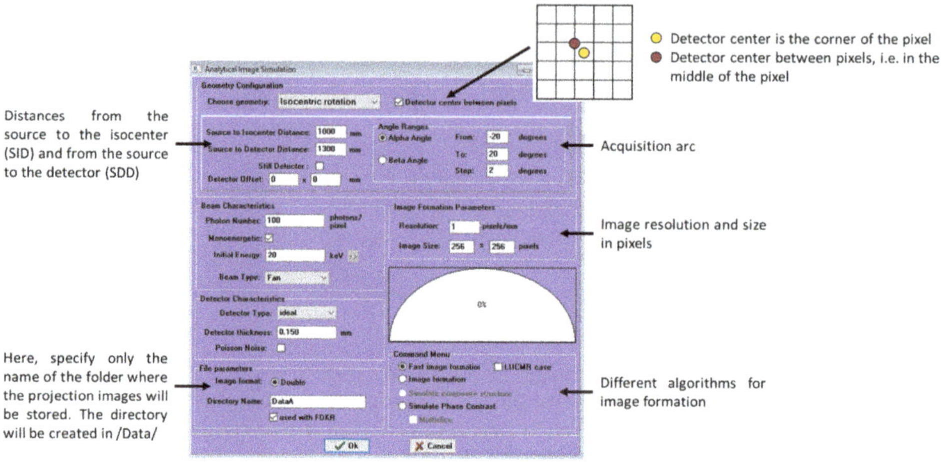

Figure 4.41. Module for image formation modelling and simulation of CSG-based phantoms.

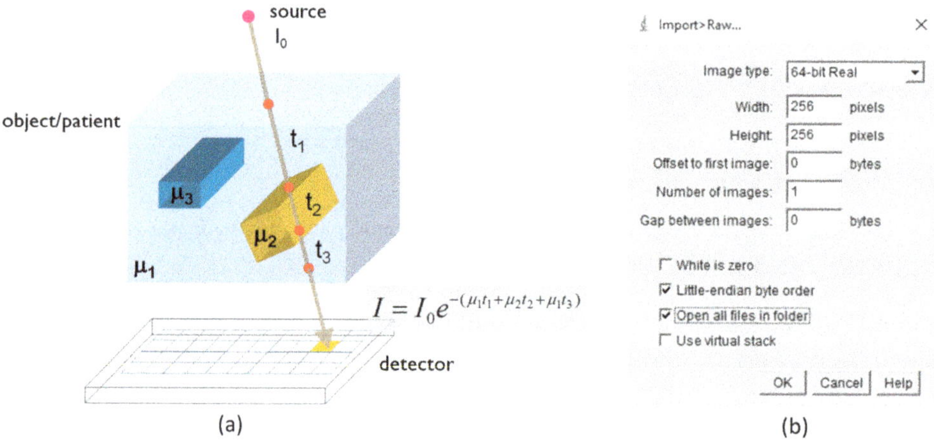

Figure 4.42. Modelling and simulation of x-ray images: (a) demonstration of ray concept and pixel value formation and (b) opening of projection images generated with the *Image Simulation—solid geometry* module.

Two approaches to implementing the Beer–Lambert law are used that differ in computing time efficiency. Figure 4.42(a) schematically shows how a pixel value is formed.

- Fast image formation approach:

$$\text{pixel value} = \mu_1(t_1 + t_2 + t_3) + \mu_2 t_2, \tag{4.9}$$

where μ_1 and μ_2 are the linear attenuation coefficients specified in the *UserAttCoeff* edit box from the *Phantom* module. In order to obtain correct simulation results, in the *Phantom* module recalculation of μ of the objects is always needed, except those that are not embedded in another object. For the case illustrated in figure 4.42(a), the first object will have μ_1, while the μ coefficients of the second and the third objects will be $(\mu_2{-}\mu_1)$ and $(\mu_3{-}\mu_1)$, respectively.

- Image formation:

$$\text{pixel value} = \mu_1 t_1 + \mu_2 t_2 + \mu_1 t_3, \tag{4.10}$$

where μ_1 and μ_2 are the linear attenuation coefficients defined by the program as a function of the energy. The modelled linear attenuations are listed in table 4.1. In case these are not defined in the database, then μ_1 and μ_2 are the linear attenuation coefficients specified in the *UserAttCoeff* edit box in the *Phantom* module.

These two approaches are reflected by the radio buttons *Fast image formation* and *Image formation*. The distances from source to detector and isocenter are specified in the panel. This panel offers the possibility to completely define the acquisition geometry.

4.6.2.1 Simulation of isocentric imaging

Modelling and simulation of isocenter of x-ray imaging (figure 4.39(a)) is demonstrated by the following example. Let us create a new solid-based phantom, composed of one cylinder with the parameters specified in table 4.7.

Use the *Image Simulation—solid geometry* option and generate one image at the 0° position of the gantry. The image size is 256 pixels × 256 pixels as each pixel is 1 mm. The distances from the source to the isocenter and the detector are 1000 mm and 1300 mm, respectively. These settings are depicted in figure 4.43(a), while the generated image is shown in figure 4.43(b). The magnified version of the x-ray image shown in figure 4.43(c) depicts the so-called voxelisation effect. The increase of a factor of 4 in the pixel resolution (pixel size = 25 μm) shown in figure 4.44(a) results in a smoother projection image of the cylinder (figure 4.44(b)). Projection images are generated in folder/*Data*/*DataA*, while the format of the images is double precision.

In order to simulate isocentric rotational imaging (figure 4.39(a)), for instance tomosynthesis implemented with the x-ray tube moving every 2 degrees in a 50 degree acquisition arc, the following action is required. In figure 4.44(a), the acquisition arc for which the images are acquired and the step −25°:2°:25° are specified. Images may be reviewed by using the approach demonstrated in figure 4.42(b). For exercise purposes repeat the same exercise for 0°:1°:359°, simulating cone-beam CT imaging.

Table 4.7. Parameters of the cylindrical object.

Object type	Centre	Dimensions	Rotation	Att. coeff.
Cylinder	$X = 0$ mm $Y = 0$ mm $Z = 0$ mm	Radius = 10 mm Height = 44 mm	$\phi = 45°$ $\theta = -72°$ $\psi = -45°$	0.078 mm^{-1}

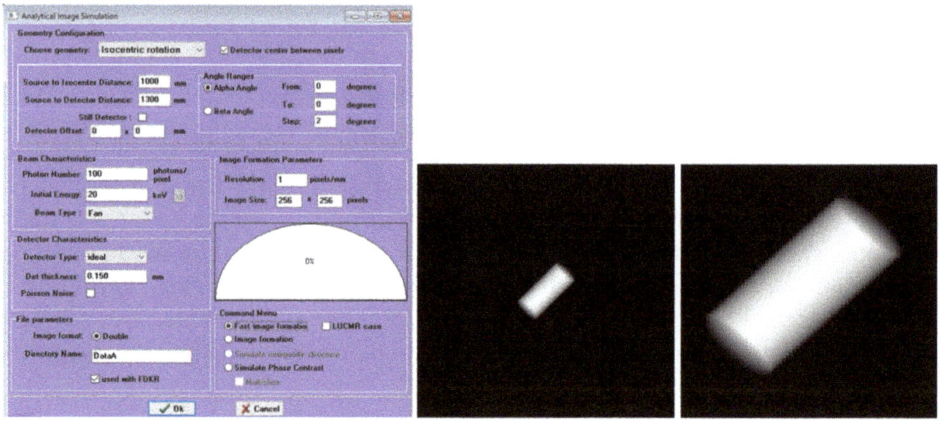

(a) (b) (c)

Figure 4.43. Image formation: (a) settings for obtaining an x-ray image of a cylinder and (b, c) generated x-ray image.

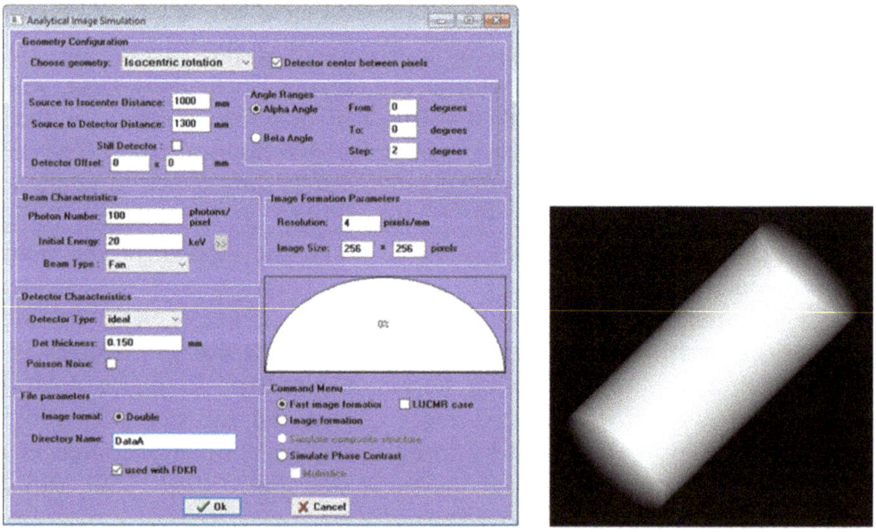

(a) (b)

Figure 4.44. Image formation: (a) settings for obtaining an x-ray image of a cylinder with pixel dimensions of 25 μm on each side and (b) simulated x-ray image of size 256 pixels × 256 pixels.

4.6.2.2 Simulation of mammography

Modelling and simulation of mammography imaging (figure 4.39(c)) is demonstrated by the following example. Let us create a new solid-based phantom, composed of one semi-cylinder with the parameters shown in table 4.8.

Mammography simulation is implemented by choosing the *Mammography* option from the combo box *Choose Geometry*. Figures 4.45(a)–(c) show images

Table 4.8. Parameters of the semi-cylindrical object.

Object type	Centre	Dimensions	Rotation	Att. coeff.
Semi-cylinder	$X = 0$ mm $Y = 0$ mm $Z = 0$ mm	Radius = 80 mm Height = 40 mm	$\phi = 135°$ $\theta = 0°$ $\psi = 45°$	0.078 mm^{-1}

(a) (b) (c)

Figure 4.45. Generated x-ray images: (a) pixel size = 1 mm, image size = 256 pixels × 256 pixels, source positioned at 0°; (b) pixel size = 0.2 mm, image size = 1024 pixels × 1024 pixels, source positioned at 0°; (c) pixel size = 0.2 mm, image size = 1024 pixels × 1024 pixels, source positioned at −26°.

generated for SID and SDD 600 mm and 650 mm, respectively. The image with a size of 256 pixels × 256 pixels in figure 4.45(a) is generated from a craniocaudal view which corresponds to 0° x-ray tube position, and a resolution of 1 mm per pixel; the image in figure 4.45(b) is generated for an x-ray tube positioned at 0°, and a resolution of 200 μm per pixel and an image size of 1024 pixels × 1024 pixels, while the image in figure 4.45(c) is generated for an x-ray tube positioned at −26° (see figure 4.39(c)) while the rest of the settings are similar to those used for the generation of the image in figure 4.45(b).

4.6.3 Tutorial with voxel anthropomorphic phantoms

The module is used to analytically simulate the x-ray transport through voxel phantoms. The ray tracing is implemented by equation (4.8), where in this case n is the number of discrete steps along the ray line from the source towards the detector pixel. Figure 4.46 presents the process of forming each voxel value in the image.

The module consists of two sub-panels: a panel for loading the voxel phantom, shown in figure 4.47(a) and a panel for simulation of image formation, shown in figure 4.47(b). As can be seen, the parameters used to describe the image formation for voxel phantoms are similar to those used to describe the image formation from solid-based objects. This module is discussed in detail in chapter 5.

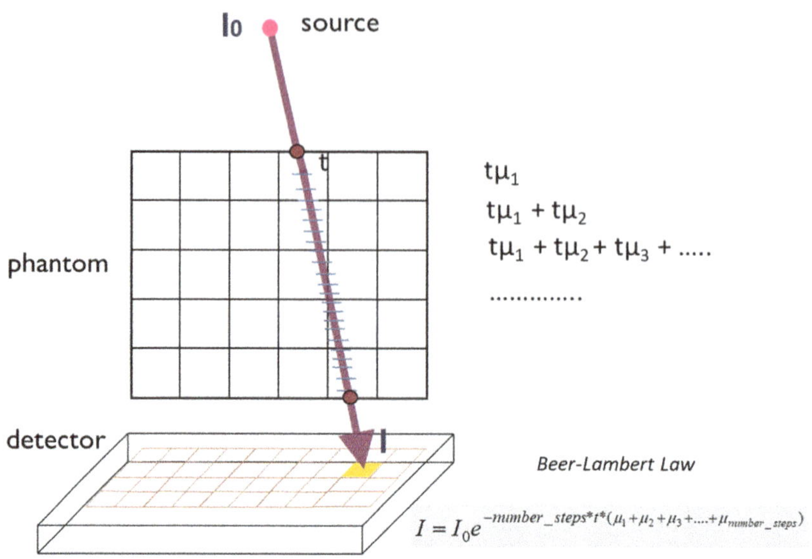

Figure 4.46. Sketch of the process of pixel value formation in the image.

(a) (b)

Figure 4.47. The module used to generate x-ray images from voxel phantoms: (a) loading the voxel phantom and (b) the parameters describing the x-ray image formation.

4.6.4 Self-training example

Exercise 1: Simulation of x-ray image formation using a solid-based phantom

The aim of this self-training exercise is to generate x-ray images of the phantom created in section 4.3.2. exercise 1 and shown in figure 4.15. Images created with *Analytical Image Simulation → Image Simulation—solid geometry* are always stored as arrays of real values of double precision. The values in these images are line integrals.

Tasks:

(a) Use the *XRAYImagingSimulator* and load the created *paraffin_phantom.phn* phantom.

(b) Use the *Analytical Image Simulation → Image Simulation—solid geometry* and tune the following settings:
   - SID: 500 mm; SDD: 540 mm.
   - Fan beam profile.
   - Energy: 18 keV.
   - Image resolution: 10 pixels/mm; image size: 700 × 700 pixels.
   - Acquisition angle: from 0° to 0° with a step of 2^0 degrees.
   - No noise added, i.e. ideal images.

(c) Generate a planar projection image of the phantom at 0° and save the image in a new folder *Excersise_4.6.4.1a.*

(d) Change the material of one of the spheres from water to PMMA. Remember the coordinates of this sphere.

(e) Use the *Analytical Image Simulation → Image Simulation—solid geometry* and the above settings to generate a similar planar image from this phantom at 0°. Save the image in a folder *Excersise_4.6.4.1b.*

(f) Visualise the two projection images and try to find the difference between them. If no differences are noticeable, try to prove that there are no differences at all.

(g) Use *Analytical Image Simulation → Image Simulation—solid geometry* and the above settings to generate 26 projection images along an arc of −25° to 25° from the second phantom (with the PMMA sphere). For SDD and SID use the above settings. Store the images in a folder *Excersise_4.6.4.1c.*

(h) Use *ImageJ* or *MATLAB* to inspect the images, the values of which are of double precision, i.e. 64 bit reals.

Exercise 2: Simulation of x-ray image formation using a solid-based phantom

The aim of this self-training exercise is to generate x-ray images of the phantom generated in section 4.4.2, exercise 2 (figure 4.23). This phantom presents a container of paraffin, filled with hundreds of water spheres. The diameter of the spheres is 8 mm, while the size of the container is 60 mm × 60 mm × 40 mm (width, length, height).

When performing image simulation of these kinds of phantoms, the *Simple Irradiation* and the option *LUCMFR case* (checked) are used together. This option forces an automatic recalculation of the attenuation coefficients of the used objects within the phantom according to the energy, specified in the *Energy* field. Images, created with *Analytical Image Simulation* → *Image Simulation—solid geometry* are always stored as arrays of real values of double precision. The values in these images are line integrals.

Tasks:

(a) Use the *XRAYImagingSimulator* and load the created **.phn* phantom.
(b) Use the *Analytical Image Simulation* → *Image Simulation—solid geometry* and tune the following settings:
  - SID: 600 mm; SDD: 650 mm.
  - Fan beam profile.
  - Energy: 20 keV.
  - Image resolution: 5 pixels/mm; image size: 450 × 450 pixels.
  - Acquisition angle: from 0° to 0° with a step of 2° degrees.
  - No noise added, i.e. ideal images.
(c) Generate a planar projection image of the phantom at 0° and save the image in new folder *Excersise_4.6.4.2a.*
(d) Change the material of one of the spheres from water to PMMA. Remember the coordinates of this sphere.
(e) Use the *Analytical Image Simulation* → *Image Simulation—solid geometry* and the above settings to generate a similar planar image from this phantom at 0°. Save the image in a folder *Excersise_4.6.4.2b.*
(f) Visualise the two projection images and try to find the difference between them. If no differences are noticeable, try to prove that there are no differences at all.

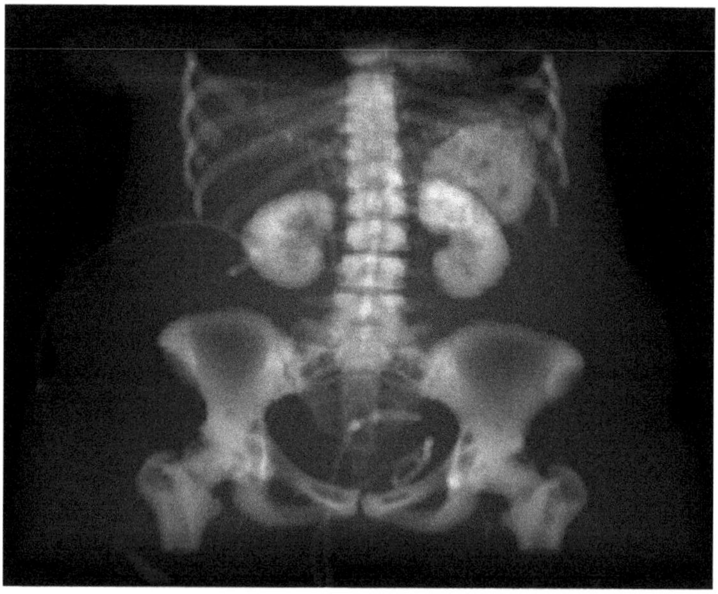

Figure 4.48. An x-ray image, obtained from a voxel phantom created from patient CT images.

(g) Use *Analytical Image Simulation → Image Simulation—solid geometry* and the above settings to generate 26 projection images along an arc of −25° to 25° from the second phantom (with the PMMA sphere). For SDD and SID use the above settings. Store the images in a folder called *Excersise_4.6.4.2c*.
(h) Use *ImageJ* or *MATLAB* to inspect the images, the values of which are of double precision, i.e. 64 bit reals.

Exercise 3: Simulation of x-ray image formation using a voxel phantom
The aim of this self-training exercise is to generate x-ray images (figure 4.48) of the phantom generated in section 4.5.1 (figure 4.26). This phantom is a liver voxel phantom based on patient data. To create an image of this phantom, use the option *Image Simulation—voxel geometry.* The values in these images are line integrals.

Tasks:

(a) Use the *XRAYImagingSimulator*.
(b) Use the *Analytical Image Simulation → Image Simulation—voxel geometry* and tune the following settings:
  - SID: 1000 mm; SDD: 1300 mm.
  - Fan beam profile.
  - Energy: 20 keV.
  - Image resolution: 5 pixels/mm; image size: 800 × 800 pixels.
  - Acquisition angle: from −0° to 0° with a step of 2° degrees.
  - No noise added, i.e. ideal images.
(c) Generate a planar projection image of the phantom at 0° and save the image in a new folder *Excersise_4.6.4.3*.
(d) Use *ImageJ* or *MATLAB* to inspect the images, the values of which are of single precision, i.e. 32 bit reals.

References

Bliznakova K *et al* 2010 Experimental validation of a radiographic simulation code using breast phantom for x-ray imaging *Comput. Biol. Med.* **40** 208–14

Bliznakova K 2016 Development of breast software phantom dedicated for research and educational purposes *Radiat. Appl. J.* **2** 14–19

Cockmartin L *et al* 2012 Design and evaluation of a phantom with structured background for digital mammography and breast tomosynthesis *Int. Workshop on Digital Mammography* Lecture Notes in Computer Science vol 7361 (New York: Springer) 642–9

Dukov N, Bliznakov Z, Buliev I and Bliznakova K 2019 Creation of computational breast phantoms with extracted abnormalities from real patient images *World Congress on Medical Physics and Biomedical Engineering 2018 IFMBE Proc.* vol 68/1 ed L Lhotska, L Sukupova, I Lacković and G Ibbott (Singapore: Springer)

Gang G J *et al* 2010 Anatomical background and generalized detectability in tomosynthesis and cone-beam CT *Med. Phys.* **37** 1948–65

Johns H E and Cunningham J R 1983 *Physics of Radiology* (Springfield, IL: Charles C Thomas)

Marziani M *et al* 2002 Dual-energy tissue cancellation in mammography with quasi-monochromatic x-rays *Phys. Med. Biol.* **47** 305–13

Shepp L A and Logan B F 1974 The Fourier reconstruction of a head section *IEEE Trans. Nucl. Sci.* **21** 21–43

Taibi A *et al* 2003 Dual-energy imaging in full-field digital mammography: a phantom study *Phys. Med. Biol.* **8** 1945–56

Yang X *et al* 2011 Automatic tissue classification for high-resolution breast CT images based on bilateral filtering *Proc. SPIE* **7962** 79623H

Chapter 5

Modelling of brcast phantoms and simulation of x-ray breast imaging techniques

5.1 Introduction

Breast cancer remains the most common cause of death for women below seventy years of age. In Europe, one in ten women will develop breast cancer in her lifetime. Early diagnosis is recognised as a critical factor that improves the chance of survival. As the population ages, there will be an increase in the number of women that will be affected by this disease. Currently, x-ray mammography screening carried out according to EU guidelines is the best form of early breast cancer detection (Olsen *et al* 2005). Despite technological advances, such as digital mammography, screening and diagnosing cancers hidden in dense breast parenchyma remains a challenging task (Schulz-Wendtland *et al* 2009). The insufficient soft tissue contrast and the effect of overlying structures in x-ray mammograms are the basic causes that result in approximately 10%–20% of palpable breast cancers being missed during screening mammography. The development of digital detector technology led to the introduction of novel advanced x-ray techniques that, applied to breast imaging, may result in improved detection and diagnosis of breast cancer. Two examples are breast tomosynthesis (BT) and breast CT, which have the potential to be more accurate in the case of dense breasts and to extract additional diagnostic information in comparison to conventional mammography. Related to this, modelling and simulation are important tools, primarily exploited in the development, optimisation, and the establishment of novel breast imaging techniques for screening and diagnostic imaging. These are instruments commonly used to evaluate both the performance of new modalities in their entirety (e.g. BT or dedicated breast CT) as well as for specific parts of the x-ray imaging systems (e.g. detectors, x-ray spectra, or acquisition geometries). The main components of breast imaging simulations involve modelling the components of the x-ray imaging system: the x-ray spectra,

doi:10.1088/2053-2563/aae197ch5

detector, acquisition geometry, and the breast itself, and a computer code to simulate the radiation transport in the model, resulting in generation of projection mammography images and evaluation of the dose absorbed in the breast. The precise modelling of all these components will ensure results approaching what would be obtained in practice.

The focus of this chapter is accurately modelling the x-ray breast imaging chain using the *BreastSimulator* software application. The aim is to provide the reader knowledge on how to create realistic breast models, to model any x-ray breast imaging technique, and to simulate image formation. Exercises are provided at the end of the chapter.

5.2 Overview of the *BreastSimulator* software tool

The *BreastSimulator* (Bliznakova *et al* 2012) is a software application dedicated to x-ray breast imaging research. It consists of four major parts which are outlined in figure 5.1(a): (a) a module for breast modelling, (b) a module for breast compression, (c) a module for image formation, and (d) a module for visualisation of images and 3D breast models. Figure 5.1(b) shows a screen shot from the main window of the software application.

The first module component of the *BreastSimulator* is *3D Breast Modelling*, which is dedicated to the generation of 3D computational breast models; the second module is *Breast Compression*, which is used to simulate breast compression. Simulation of synthetic images in different acquisition geometries is implemented in the *Image Simulation* module where the user can model any image acquisition geometry and generate projection images. Finally, there are two modules which facilitate the visualisation of both computational 3D breast models and synthetic images.

The *BreastSimulator* software package can be run on Windows. It is compiled with *Embarcadero C++ Builder* as a single application with a graphical user interface (GUI) that combines all four modules. The executable for Windows,

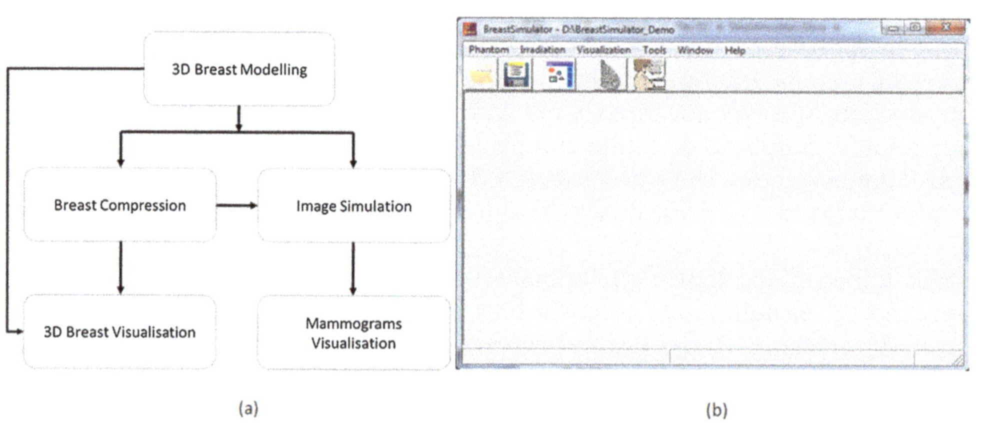

Figure 5.1. *BreastSimulator* software application: (a) major functionalities and (b) the main window of the application.

with relevant help and demo files, will be available for the participants in the EUTEMPE course. Issues related to the installation of the software application *BreastSimulator* can be found in the *SoftwareRequirements* text file in the supplementary material of the chapter.

5.3 Breast anatomy and imaging

5.3.1 Anatomy of the normal breast

The adult female non-lactating breast, shown in figure 5.2, is composed of the following structures. The mammary gland is located within the superficial fascia of the anterior chest wall and is covered by the skin. It consists of glandular tissue of the tubuloalveolar type (parenchyma), connective tissue (stroma), and other non-glandular tissue types (lymphatics, adipose tissue, blood vessels, and nerves). The mammary gland consists of 15–20 irregular-pyramidal shaped lobes. Each lobe is drained by its own lactiferous duct, which is partitioned into narrowed and shorter branches, thus it has a tree structure. At the suboreolar area the lactiferous ducts dilate to form the lactiferous sinuses (ampullae). At the ampullae, ducts of several lobes merge to form major ducts that emerge in the nipple independently, forming separate orifices on the nipple's surface. There are 6–8 orifices draining to the nipple

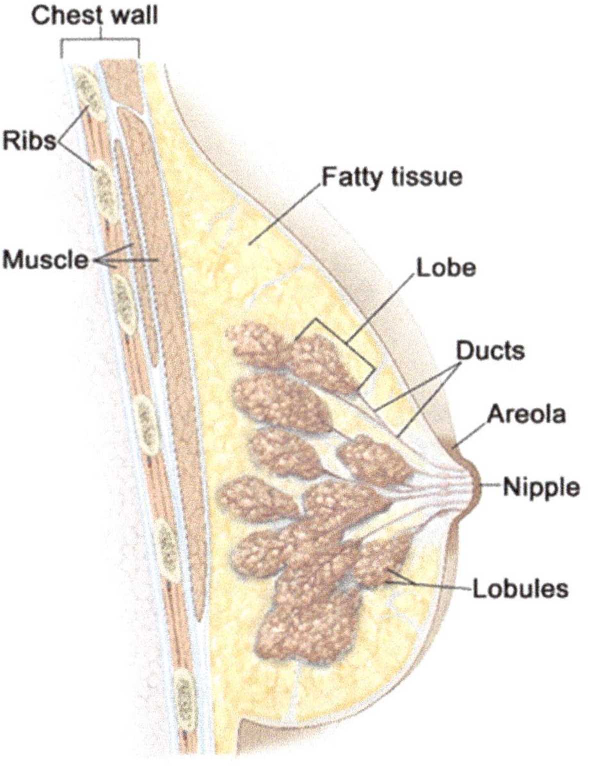

Figure 5.2. Anatomy of the female breast. (Source: https://www.teresewinslow.com/#/breast/.)

(Egan 1988). In the adult female breast each lobe consists of several lobules. A lobule, 1 mm to 2 mm in diameter (De Paredes 1992), is composed of ductules and alveoli that end at one common terminal duct. This structure, called a *terminal duct-lobular unit* (TDLU), is located at the ends of the tree branches. Both ducts and lobules are embedded in loose connective tissue and are separated by dense connective tissue.

Adipose and glandular tissues are the two predominant tissue types within the breast. As the mammary gland is located within the superficial fascia, it is separated from the skin by 0.5 cm to 2.5 cm of adipose tissue (subcutaneous fat). The ducts before the ampullae measure 1–2 mm in diameter while the major ducts' diameters vary between 2 mm and 4.5 mm (Egan 1988). The total length of the ducts varies from 2 to 4.5 cm (Junqueira *et al* 1989). Cooper ligaments are bands of dense connective tissue that support the breast, connecting the dermis of the skin with the deep fascia. They constitute fingerstall-like compartments. The lobes of the gland tissue and the fat are situated in these compartments (Novak 1988).

5.3.2 Anatomy of breast abnormalities

Two major types of abnormalities are considered on a mammogram: masses and calcifications (De Paredes 1992). *Masses* are groupings of abnormal cells; their most important characteristics are listed in table 5.1.

Table 5.1. Breast mass characteristics.

Feature	Short description
Location	Masses are detected everywhere in the breast.
Size	1–4 cm
Shape	The shape of a mass can be:  (i) round (ii) oval (iii) lobular (iv) irregular
Margins	The margin is the border of the mass and its analysis is one of the most important criteria in determining whether the mass is likely to be benign or malignant. Margins can be:  (i) circumscribed (ii) microlobulated (iii) obscured (iv) indistinct (v) spiculated
Density	The x-ray attenuation coefficient of the mass.
Calcifications	When found within a mass, they provide information on the type of mass.
Examples	Fibroadenoma, cysts, intraductal pappiloma, duct hyperplasia, abscess, cystosarcoma phylloides

Calcifications are tiny mineral deposits produced by the cells as a result of some benign or malignant processes, with sizes of 0.1 mm to 1 cm. The presence of calcifications in the breast tissue is one of the main features considered by radiologists. Their characteristics are listed in table 5.2.

Both abnormalities are subjected to various classification schemes, whose final scope is to differentiate them as being malignant or benign (Zonderland and Smithuis 2013). They vary greatly in their size, shape, location, density, margins, number, and distribution. None of these parameters alone is used to identify the malignancy or benignity of the abnormalities; rather a complete analysis of a set of relations between these parameters must be accomplished. Round, oval, or lobulated *masses* with sharp and well-circumscribed borders, dimensions less than 1.5 cm, and that contain low-density regions of fat or are entirely fatty, are most likely to be benign. Malignant *masses* possess high-density, spiculated margins and indistinct, irregular, or fuzzy contours. Benign *calcifications* are usually scattered in distribution and have uniform shape and density. The *macro-calcifications* are usually large and round and almost always recognised to be benign. *Malignant calcifications* are more irregular in shape, size, and distribution. They are generally smaller than benign calcifications and are of particular clinical significance when found in clusters of five or more in a 1 cm^2 area (Sickles 1986). Calcifications outside the glandular tissue are not a sign of breast cancer.

5.3.3 Breast imaging

In *mammography*, the normal anatomy projected on a mammogram includes several or all of the following breast components: pectoralis muscle, pectoralis fascia, axillary lymph nodes, arteries, veins, retromammary fat, Cooper ligaments, glandular tissue, major and lactiferous ducts, nipple, areola, and skin. The pectoralis muscle and fascia as well as the axillary lymph nodes are visible in the routine

Table 5.2. Breast calcification characteristics.

Feature	Description
Location	They can be found within and around the (i) ducts, (ii) lobules, (iii) vascular structures, (iv) skin, (v) interlobular connective tissue or fat, and (vi) masses.
Size	(i) Macrocalcifications—calcifications with a size of 2.0 mm or larger; (ii) microcalcifications—calcifications with sizes less than 1 mm.
Number	This parameter is very important for the calcifications found to form a cluster and is used to classify the cluster as benign or malignant.
Morphology	Shape and margins (the same as the masses).
ACR classification	(i) Benign calcifications: skin; milk-of-calcium; round; popcorn-like; punctuate; rim or egg-shell; vascular; rod-shaped; suture; dystrophic; spherical or lucent-centred. (ii) Indeterminate calcifications: indistinct; amorphous. (iii) Higher probability of malignancy: heterogeneous (granular); fine, linear, branching (casting).

oblique view and appear as a white-bright area. Lactiferous ducts, unless enhanced by a contrast agent, are invisible or barely visible, because of the surrounding connective tissue. Major ducts can be seen to a limited extent in the subareolar region. Instead, a dense triangular fan-shaped density is presented beneath the areola. Vascular structures appear as smooth and undulating structures.

Darker regions on a mammogram represent the projection of fatty tissue, due to its radiolucency, while the fibroglandular and connective tissues project as medium-to-high bright regions due to their increased x-ray absorption. The skin attenuates the x-rays significantly; however due to its small thickness it appears only slightly visible. Cooper ligaments can be depicted as fine curved lines with low intensity. All other anatomical structures appear as shades of white. Breast masses, including both non-cancerous and cancerous types, appear as bright regions. Calcifications are seen as very small spots that are relatively brighter than the surrounding tissue.

In addition to mammography, there are some other advanced x-ray imaging techniques used to screen and diagnose the breast. *Breast tomosynthesis* is a pseudo-three-dimensional x-ray imaging technique of the breast that reconstructs tomographic images from a set of angular projections with dose levels not exceeding those of two conventional mammograms acquired in an arc around the patient breast. The acquisition geometry in breast tomosynthesis is very similar to that used in mammography. The difference is in the rotation of the x-ray tube around the breast, which is compressed, and a series of images is acquired, one at each x-ray tube position. During the acquisition process, the detector can either be static or rotate to maintain its top surface normal to the x-ray tube. An excellent review on breast tomosynthesis can be found in Sechopoulos (2013). In *breast CT*, 300–500 low-dose projections are acquired in a circular trajectory around the breast using a flat panel detector, followed by image reconstruction to provide the 3D breast volume (O'Connell *et al* 2014). The average glandular dose to the breast from breast CT can range from as little as a two-view screening mammogram to approximately that of a diagnostic mammography examination. Breast CT displays 3D images of the internal structures of the breast; therefore, evaluation of suspicious features such as microcalcifications, masses, and asymmetries can be made in multiple anatomical planes from a single scan (O'Connell *et al* 2014).

5.4 Breast model generation

The complete breast model is an aggregate (shown in figure 5.3) of breast shape, duct system, Cooper ligaments, texture, and abnormalities. First, modelling of the breast's external shape is performed. Ducts are then generated starting from the nipple and restricted by the external breast contour. The 3D background texture is mapped to the volume, defined by the breast external surface. The composition process continues with adding Cooper ligaments and the pectoralis muscle. Finally, modelled abnormalities are inserted in order to complete the breast phantom.

To create a computational breast phantom, the *Compose Breast* option is activated from the main menu, shown in figure 5.4. The screen *Breast Modelling* is divided in three parts: *Breast Design, Generated Breast Data*, and *Parameters*. The

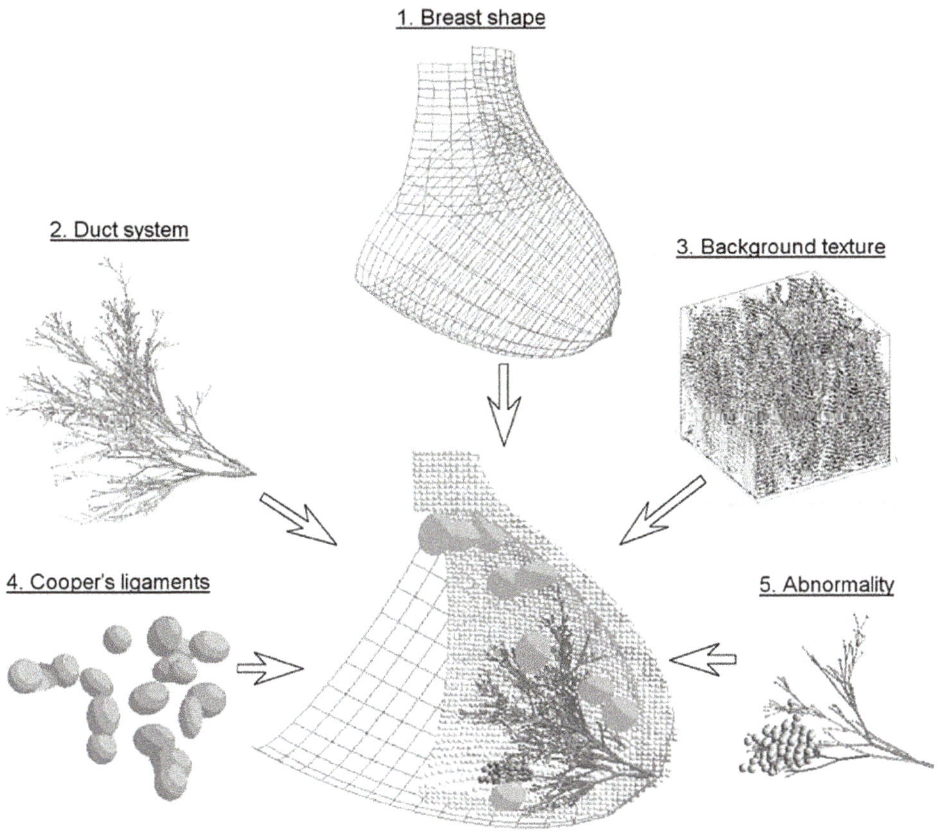

Figure 5.3. Steps in the process of composing the breast. (Reproduced from Bliznakova *et al* 2003 with the permission of the publisher.)

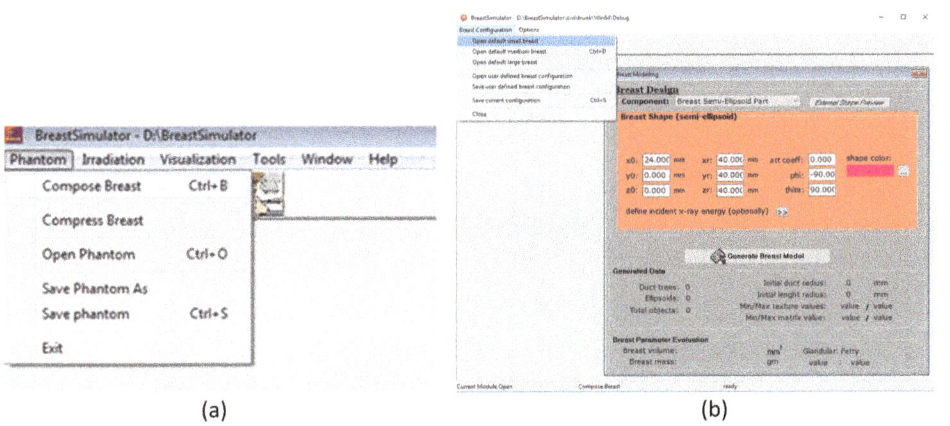

Figure 5.4. Creation of a computational breast model: (a) the *Compose Breast* option used to create a new breast phantom and (b) the panel which appears as a consequence of *Compose Breast* action.

breast model is created as a sequence of modelling the different breast parts, shown in figure 5.3.

The *Breast Design* panel contains of a set of changing panels used to model the individual breast components. This panel consists of several sub-modules that are used to model the different breast components: (a) external shape; (b) glandular tissue: ducts and Cooper ligaments; (c) adipose tissue; (d) abnormalities such as regular and irregular masses, spherical and ellipsoid calcifications, and groups of microcalcifications (μCs), elongated abnormalities; (e) skin, pectoralis, and lymphatics.

Briefly described, the breast surface is modelled as a combination of two geometrical primitives: an elongated semi-ellipsoid and an elongated semi-hyperboloid. For breast CT and cone-beam CT applications the external shape may be modelled with a single semi-ellipsoid. The duct system is simulated using a network of cylinders, probabilistically arranged in the breast as branches, in a tree-like arrangement, starting from the nipple and restricted by the external breast contour. The mammographic texture simulates the presence of adipose, fibrous, and connective tissues, as well as other non-glandular tissue types that are not explicitly modelled. The algorithm is based on the use of random walks, calculated using the concept of the 'fractional Brownian motion model', followed by a series of 3D image processing algorithms that result in a realistic 3D texture. Cooper ligaments are modelled as a set of thin ellipsoid shells, originating at randomly sampled positions in the breast model, while the pectoralis muscle is approximated as a cone-shaped object. Breast abnormalities are modelled with round, ovoid, elongated, or irregular shapes.

5.4.1 Breast external shape

The breast shape, shown in figure 5.5(a), is modelled as a combination of two geometrical primitives: an elongated semi-ellipsoid (figure 5.5(b)) and an elongated semi-hyperboloid (figure 5.5(c)) that resembles the breast shape. The variation in the geometrical parameters defining these two primitives allows the creation of breast objects of different sizes and shapes.

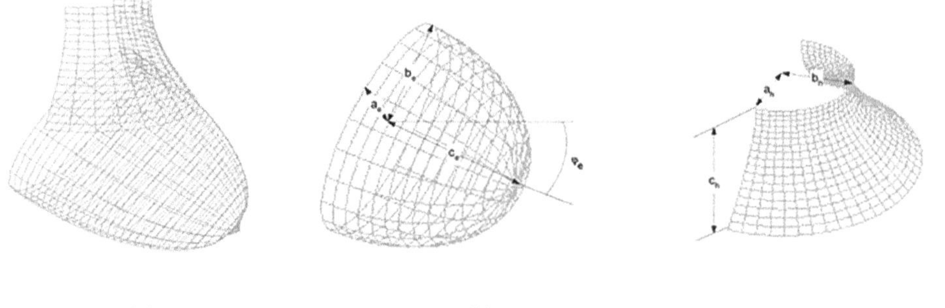

Figure 5.5. Breast design: (a) design of the breast shape which is a combination of (b) the semi-ellipsoid part and (c) semi-hyperboloid part. (Reproduced from Bliznakova *et al* 2003 with the permission of the publisher.)

The *breast semi-ellipsoid* is described with the semi-axes of the semi-ellipsoid, its location in the 3D scene, and the rotation. Default values of semi-ellipsoid axes for small, medium, and large breast shapes are shown in figure 5.6(a). To implement this shape, the software tool has an option for modelling of this shape: *Breast Semi-Ellipsoid Part* (figure 5.6(b)). This panel describes the semi-ellipsoid object: (x_0, y_0, z_0), (*xr*, *yr*, *zr*), and (*phi*, *thita*) are the position coordinates, the semi-axes of the semi-ellipsoid object, and two of its Euler angles, respectively. The third Euler angle, *psi*, is set to zero. This solid object is assigned a linear attenuation coefficient μ of zero. During the breast creation, the attenuation coefficients generated in the texture (background) matrix are mapped to this object.

The *Breast Hyperboloid Part* (figure 5.7(a)) is a semi-hyperboloid, defined in table 4.2, chapter 4. Figure 5.7(b) shows the implementation of the geometrical primitive. In this implementation, (x_0, y_0, z_0), (*xr*, *yr*, *zr*), and (*phi*, *thita*) are the position, the semi-axes of the object, and two of its Euler angles, respectively, while the third Euler angle, *psi*, is set to zero. Similarly to the semi-ellipsoid object, the linear attenuation coefficient is set to zero and the attenuation coefficients from the generated background matrix are mapped to this solid object.

The *exclude* option is used in the case when the semi-hyperboloid is not desired in the final breast model. For example, for testing a cone-beam breast CT technique, the option 'exclude' should be selected.

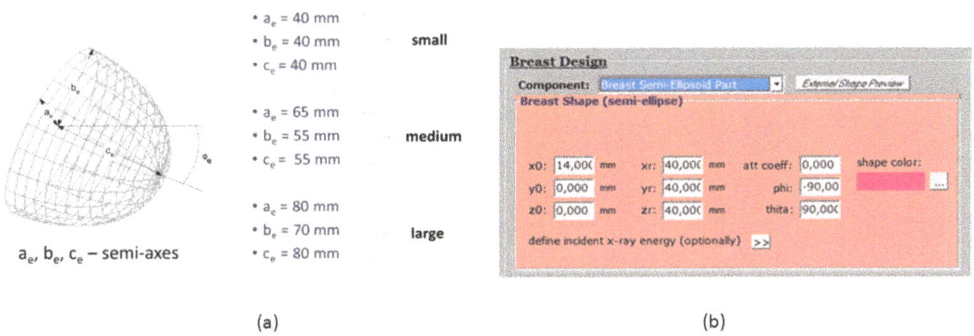

Figure 5.6. Breast design: (a) semi-ellipsoid part and (b) implementation in the program.

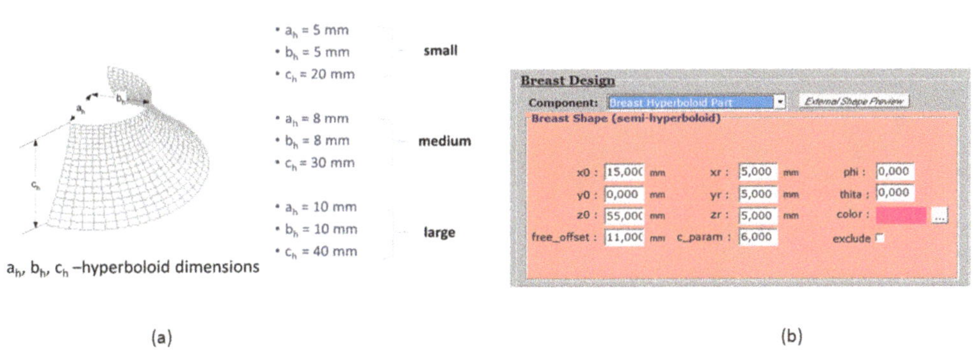

Figure 5.7. Custom hyperboloid object: (a) default values and (b) implementation in the program.

The *nipple part* is modelled as a semi-ellipsoid, shown in figure 5.8(a), similar to the breast semi-ellipsoid characteristics. The implementation is shown in figure 5.8(b).

5.4.2 Duct system

The duct system is composed of major ducts, lactiferous ducts, and lobules (figure 5.9(a)). Ducts are simulated as cylinders (figure 5.9(b)), while the lobes are simulated as spheres. The major ducts start from the breast nipple following an automated method for uniform distribution (figure 5.9(c)).

A basic glandular unit is the lactiferous duct, terminating in TDLUs. The duct system is represented by a network of cylinders, probabilistically arranged in the breast as branches, in a tree-like arrangement. The duct model input parameters include initial and final height (h_0, h_{end}), radius r_0, direction (θ_0, φ_0), and attenuation coefficient (μ_{fibrous}). Lactiferous ducts start from the end of major ducts as shown in figure 5.9(a).

The number of branches that arise from each node is randomly sampled. The branch probabilities determine the way the node will bifurcate: into two pairs of symmetric branches and/or a single branch. There is a maximum of five branches

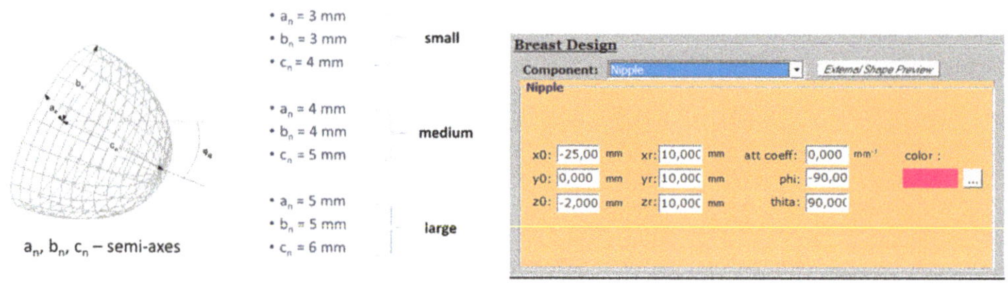

(a) (b)

Figure 5.8. Breast nipple: (a) design and (b) implementation in the program.

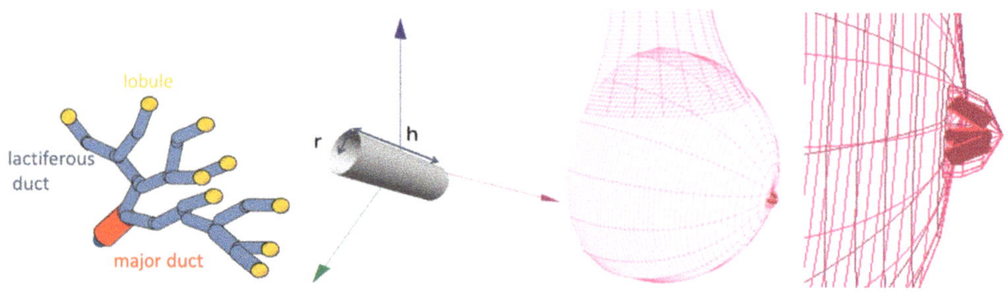

(a) (b) (c)

Figure 5.9. Construction of the duct system: (a) major duct system, (b) structure of the lactiferous tree starting from one major duct and (c) distribution of major ducts in the nipple.

that are defined in this way. A variation, introduced in these probabilities, controls the structure of the tree.

The direction of each duct branch is defined by azimuthal and polar angels (θ_i, φ_i) in a spherical coordinate system. In the program these angles are sampled as follows:

$$\varphi_i = \varphi_{i-1} + u * 2\pi \quad \theta_i = \frac{\theta_{i-1}}{2} + u * \theta_{i-1}, \tag{5.1}$$

where u is a value, uniformly distributed in [0, 1]. The 3D coordinates of each new branch node are calculated with respect to the previous one. The initial (θ_i, φ_i) are (θ_0, φ_0). The initial and the end branch nodes are joined then by cylinders with radius r_i and height h_i. The radius and height of each new duct part are calculated as follows:

$$h_i = \frac{h_{i-1}}{2}(i + rand) \quad r_i = r_0 h_i \; rand{=}[0 : 1], \tag{5.2}$$

where *rand* is a value, uniformly distributed in [0, 1].

The duct tree grows recursively with respect to each node. The growing process is limited by the breast geometry and the limitation imposed for the branch height (h_{end}). Every terminal branch ends in a sphere, simulating the lobule. The linear attenuation coefficients of the participating parts in the duct system are the same.

To model the whole duct system, the *Duct system* option is used (figure 5.10). Since the duct system starts from the nipple, the coordinates (x_0, y_0, z_0) of the duct system must coincide with those of the nipple. Major ducts are characterised by: number, height, and radius. Lactiferous ducts possess one more parameter, *limit height*, used as a criterion to limit the height of the duct. Examples for duct configurations are listed in table 5.3, while default values for major ducts, lactiferous ducts, and lobules are listed in table 5.4.

5.4.3 Breast abnormalities

Implemented algorithms for the creation of computer models of breast abnormalities result in the generation of both regular and irregular masses, calcifications, groups of microcalcifications, and elongated calcifications.

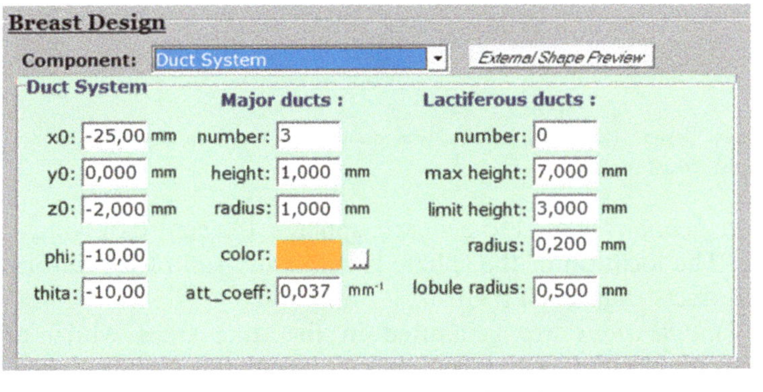

Figure 5.10. Breast design: design of the duct tree.

Table 5.3. Examples for duct configurations for a small breast model.

Major ducts	Lactiferous ducts, number	φ,°	θ,°	h_{max}, mm	h_{min}, mm	Duct radius, mm	Lobule radius, mm
3	4	−10	0	9	2	0.3	0.5
3	5	−10	0	9	3	0.3	0.5
4	5	0	0	9	4	0.1	0.5
3	4	−10	0	9	4	0.1	0.5
3	7	−10	−40	9	3	0.1	0.5
3	7	−10	40	10	3	0.1	0.5
3	7	0	−40	10	2	0.1	0.5

Table 5.4. Default values for modelling the duct system.

Object	Parameter	Value, mm
Major ducts	Height	1
	Radius	2
Lactiferous ducts	Height, h_0	7
	Height, h_{end}	3
	Duct radius, r_0	1
	Polar angle, θ_0	30°
	Azimuthal angle, φ_0	60°
Lobule	Radius	0.5

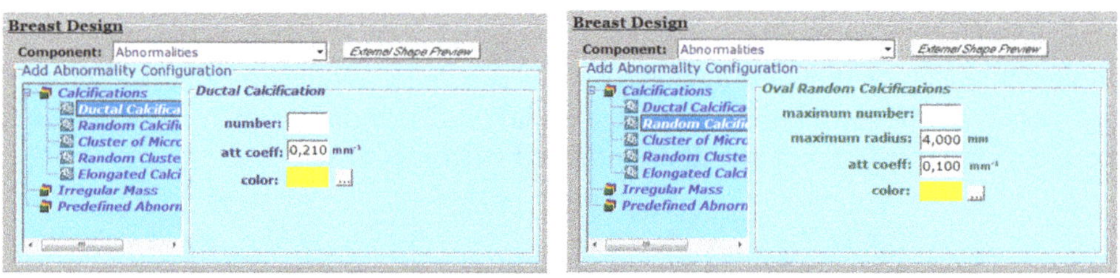

(a) (b)

Figure 5.11. Breast design: (a) ductal calcification design and (b) a randomly created cluster of micro-calcifications in the breast model.

Ductal Calcifications is used to generate spheres, the centre of which are placed in the duct tree. The location of the sphere is randomly sampled (uniformly) amongst the generated ducts (figure 5.11(a)).

Random Calcifications are generated in the duct trees which are randomly selected. The parameter *maximum number* indicates the maximal number of calcifications which may be generated (figure 5.11(b)).

Cluster of Microcalcifications creates a number of clusters of microcalcifications with distribution of a number of small calcifications, as defined in figure 5.12(a). The cluster is always associated with a duct tree. The following parameters are used to model clusters of microcalcifications. *No clusters* and *cluster radius* show the number of simulated clusters of microcalcifications and the cluster radius, respectively. Microcalcifications inside the cluster are characterised by *No of microcalc/cluster* and *microc radius range,* that is randomly sampled within the range between *Min* and *Max*, which are values reflecting the minimal and maximal radiuses of μCs to be created.

Random Cluster of Microcalcifications models user-defined *number clusters* of microcalcifications randomly placed in the surface of the breast module (figure 5.13(b)).

Elongated Calcification is used to generate a user-defined number of elongated calcifications with an attenuation coefficient *att coeff*. The parameters θ and φ are the initial Euler angles used for sampling the direction of the cylinders. The parameter *height* is the initial cylinder height used in the sampling procedure (figure 5.13a).

Irregular Mass creates abnormalities with irregular shapes (figure 5.13b). The *Irregular Mass* panel contains the following parameters used to generate irregular masses:

(a) *number*—the number of masses generated;
(b) *mass size*—the dimension of the cube inside which the mass is generated using a number of steps and iterations as described in Bliznakova *et al* 2003.

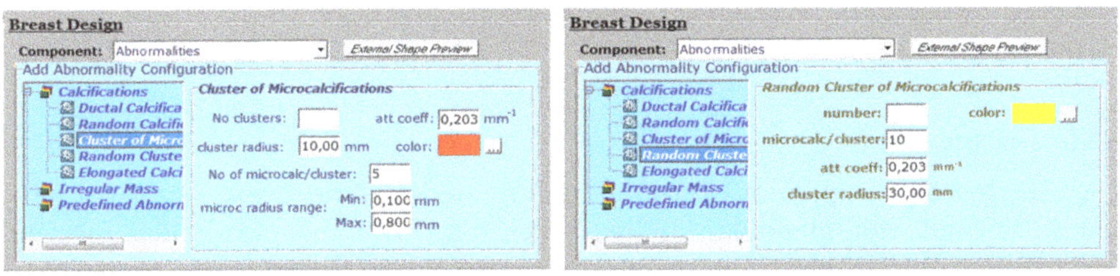

(a) (b)

Figure 5.12. Breast design: a cluster of microcalcifications (a) generated within the duct and (b) generated randomly in the phantom.

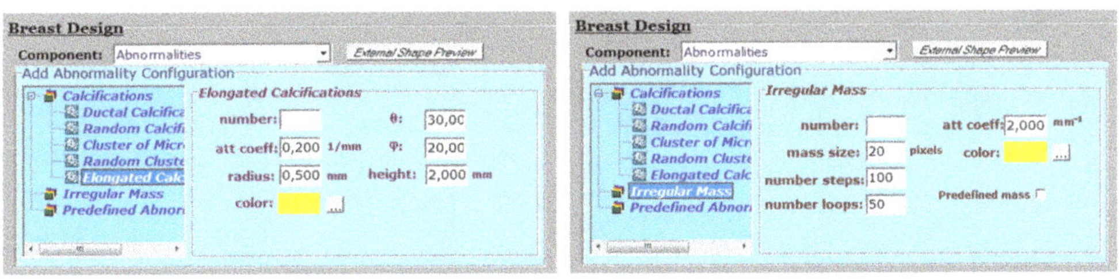

(a) (b)

Figure 5.13. Breast design: (a) an elongated abnormality and (b) an abnormality with an irregular shape.

For visualisation purposes, the mass abnormality is converted into a set of spherical objects that have attenuation coefficients set to zero. The irregular mass is introduced in the final breast matrix. The size of this mass is (mass size)^3 and is saved in *…/BreastSimulator/Phantoms* with a name *irregular_abnormality.bin.*

An abnormality may be defined separately and subsequently integrated into the breast model using the *Predefined mass* check box. This invokes a panel used to browse and load the abnormality created in advance. The latter is saved in a single precision format.

5.4.4 Cooper ligaments, pectoralis, and skin

Cooper ligaments are modelled as thin ellipsoid shells. The algorithm populates these ellipsoid shells at randomly sampled positions in the breast model. Their linear attenuation coefficient is equal to that of the ducts, while the compartments enclosed by them are assigned the attenuation coefficient of the adipose tissue (figure 5.14(a)). In addition, the generated ellipsoid shells are placed in an oblique direction (figure 5.14(b)). The application randomly samples the position of the Cooper ligaments in the breast volume (figure 5.14(c)). The generated total number of Cooper ligaments is less than the value defined by the user. This is due to the fact that some Cooper ligaments are generated outside the breast volume and therefore they are not accepted.

The *pectoralis muscle* is approximated as a cone-shaped object with a linear attenuation ($\mu_{\text{pectoralis}}$) (Johns and Cunningham 1983). Pectoralis and skin are modelled by using the *Additional Components* panel (figure 5.14(c)). *Skin* is defined by the program code.

5.4.5 Texture

The 3D mammographic background texture reflects the presence of all other glandular and non-glandular tissue types that are not explicitly modelled (Bliznakova *et al* 2010). It is represented by a voxel matrix, the dimensions of

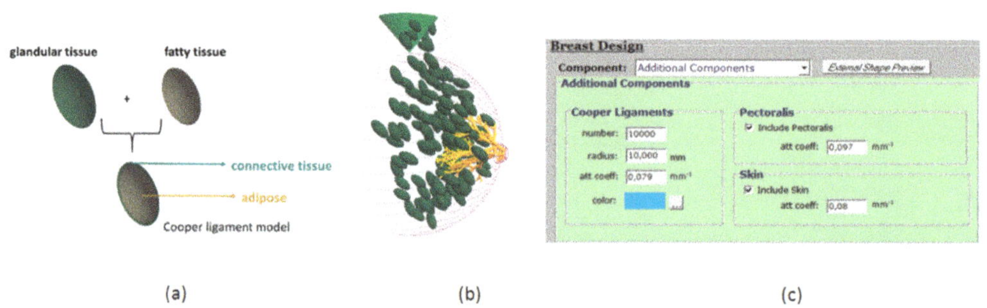

Figure 5.14. Constructing the Cooper ligaments: (a) basic elements, (b) generated in the volume, and (c) implementation in the program. Attenuation coefficients are in mm^{-1}.

which are defined by the dimensions of the external breast shape. The matrix and voxel resolutions are user selectable. The process of matrix filling strongly depends on the external breast shape. A diagram, showing the steps in the process of texture composition, is shown in figure 5.15. Four basic steps are performed for composition of the background matrix.

Initially, the 3D background matrix is filled with a voxel container using the fundamental concept of the 'fractional Brownian motion model' (fBm). This was prompted by the fact that fBm has a Gaussian nature that allows long-term correlations and has strong scaling properties, features observed in most real data and reported in the literature from detailed mammo-texture analyses. The easy implementation and simplicity of fBm has also been an advantage of using this algorithm. Various models of fBm generation have been reported in the literature, these include Poisson faulting, the Fourier spectrum, midpoint displacement, successive random additions, and summing band-limited noises. The approach followed to generate fBm is related to an algorithm called successive random additions. This approach has four basic advantages: simplicity, flexibility (for implementation and control), user-friendliness, and ease of implementation. Three one-dimensional fBm random walks (for the three dimensions) start at a voxel with a position that is sampled randomly around the nipple origin and anywhere in the breast matrix, thus forming a three-dimensional random walk. Further on, this 3D random walk moves to a neighbouring voxel that has a position, calculated from the voxel positions, determined in the previous stage as they are first displaced by addition of a random variable with a Gaussian distribution with standard deviation (σ) 1 and mean 0 (GD(1,0)),

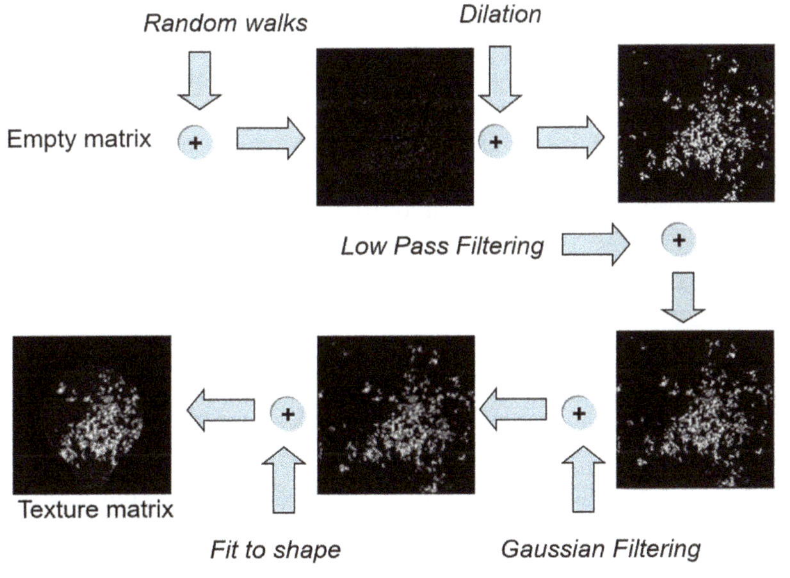

Figure 5.15. Major steps in the creation of background texture.

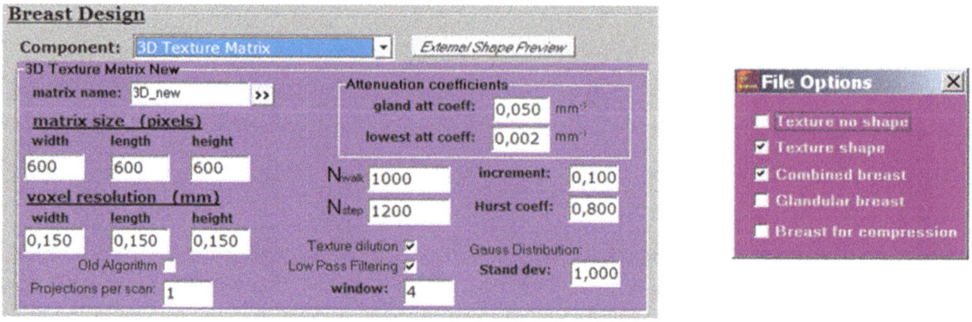

(a) (b)

Figure 5.16. Breast design—texture design: (a) the texture panel and (b) options for saving the data.

$$\begin{aligned} X_{\text{next}} &= X_{\text{current}} + \sigma^{*}\text{GD}(1, 0) \\ Y_{\text{next}} &= Y_{\text{current}} + \sigma^{*}\text{GD}(1, 0) \ , \sigma = h^{H}, \\ Z_{\text{next}} &= Z_{\text{current}} + \sigma^{*}\text{GD}(1, 0) \end{aligned} \tag{5.3}$$

where H is the Hurst coefficient and h is the increment.

The user assigns the number of steps for each walk N_{step} as well as the number of random walks N_{walk}, depending on what breast density is desired (figure 5.16a). The process stops either at the matrix boundary or when the user-assigned number of random walks is reached, N_{walk}.

Further on, a customised dilation is applied to the 3D walks. The voxels having values less than 1 obtain values, sampled based on the following equation:

$$BM(x, y, z) = \frac{k}{BM(x, y, z) + k}k + \frac{BM(x, y, z)}{BM(x, y, z) + k}BM(x, y, z), \tag{5.4}$$

where BM is the background matrix, $k = rand(0,1)$ and has a value of 0.26.

The third step is applying a user-defined low-pass filter with a value of 2 in the central voxel, while the rest of the mask voxels are set to 1. More smoothing is achievable by applying a 3D Gaussian filtering. Smoother and more blurred breast slices will be generated with higher standard deviation;

$$BM(x, y, z = \text{const}) = \frac{1}{2 * \pi * \sigma^2} \exp\left(-\frac{(x^2 + y^2)}{2 * \sigma^2}\right). \tag{5.5}$$

Finally, the 3D background texture is mapped to the volume, defined by the breast external surface. Before that, a normalisation of the values in the generated 3D texture is accomplished with respect to the maximum value that has been assigned a value of the glandular tissue. The ratio of the glandular to the fatty tissue is also calculated at the end of the composition process, as the number of glandular tissue voxels to fatty ones.

According to the algorithm discussed above, a low value of the Hurst parameter implies a rougher fBm image (fatty breast and mammogram), while H close to 1 results in a smoother fBm image (a denser breast and mammogram). Table 5.5

Table 5.5. Examples of configurations for 3D mammography texture.

	Matrix size, voxels	Voxel resolution, mm	Att coeff, cm^{-1}	N_{walk}	N_{step}	Window	Hurst coeff, H	Increment
1	256	0.3	[0.02:0.04]	100	2000	5	0.8	0.3
2	256	0.3	[0.02:0.04]	100	4000	3	0.8	0.3
3	300	0.4	[0.02:0.04]	100	1000	5	0.8	1
4	400	0.3	[0.02:0.03]	100	3000	5	0.8	1
6	600	0.15	[0.02:0.5]	1000	1200	4	0.8	0.1
7	1000	0.1	[0.02:0.5]	300	30 000	3	0.9	0.1
8	1200	0.15	[0.02:0.5]	1000	150×10^3	4	0.9	0.1
9	1500	0.15	[0.02:0.5]	10^3	200×10^3	3	0.9	0.1

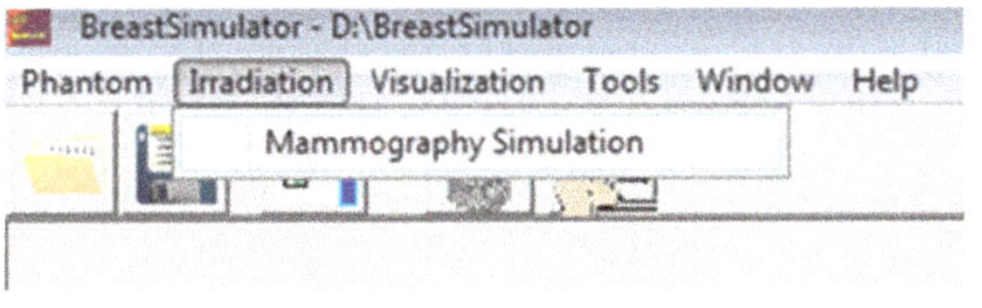

Figure 5.17. Menu option used to open the *Mammography Simulation* panel.

summarises several configurations used in the generation of different mammographic background matrices, derived from an extensive experimental work.

There are five options for saving output matrices (shown in figure 5.16(b)): (i) *Texture no shape*—saves only the generated 3D texture, not mapped to the external breast shape; (ii) *Texture shape*—saves only the texture mapped to the breast shape; (iii) *Combined breast*—combines the breast tree, simulated abnormalities, and texture, and saves the combined matrix, (iv) *Glandular breast*—saves a matrix with the percentage values of glandularity, and (v) *Breast for compression*—converts the breast values to Hounsfield units suitable for breast compression and Monte Carlo simulations.

These matrices are saved in *BreastSimulator/Phantoms/Breast/TextureMatrices* with a name specified in the edit box of the *3D Texture* panel and an appropriate suffix. The third model is useful for generation of mammographic images of the breast using monochromatic incident x-ray beams (e.g. 19 keV). The fifth model is exploited with Monte Carlo simulations and breast compression. The breast without mammographic texture is useful for the creation of breast models with different backgrounds and the same breast components comprised of 3D geometrical primitives such as duct systems, breast lesions, and so on. The texture itself may be used for the evaluation of texture characteristics and for the creation of models that have the same texture and different breast structures.

5.4.6 Breast configurations

The menu *Breast Configuration* is used to open and save configurations that contain values for the generation of a specific breast model figure 5.4(b). There are options to open default small, medium, and large breast models, as well as *user-defined breast configurations*. The user configurations are saved as binary files with the extension **.bin*. These files contain specific values for the breast model that is (to be) created: external shape dimensions; position; attenuation; number of ducts as well as their dimensions; abnormalities, etc. The default directory, where the configuration files are stored, is *..../Phantoms/Breast/Configurations*. Tables 5.6–5.8 summarise default values for small, medium, and large breasts.

Finally, the menu *Options* contains several sub-items (figure 5.4(b)): (a) *Evaluate glandularity of the whole breast* is used to evaluate the fatty to glandular ratio in the composed breast; (b) *Extract Objects* is used to save in a 3D matrix the generated solid geometrical primitives (e.g. duct tree, abnormalities, Cooper ligaments, etc) which are part of the breast model (texture is not included); and (c) *Current Process History*—shows/hides the history of the breast model generation.

Table 5.6. Default values for small breasts.

Breast shape	Number	Location xc, yc, zc, mm	Dimensions xr, yr, zr, mm	Angles, in degrees ϕ, θ, ψ	Att coeff, mm^{-1}	Colour
Breast semi-ellipsoid	1	14, 0, 0	40, 40, 40	−90°, 90°, 0°	0	Fuchsia
Breast hyperboloid	1	15, 0, 55	5, 5, 5, 6	0°, 0°, 0°	0	Fuchsia
Breast nipple	1	−25, 0, −2	2.5, 3, 3	−90°, 90°, 0°	0	Fuchsia
Duct tree—major ducts	3	−25, 0, −2		−10°, −10°, 0°	0.037	Green

Table 5.7. Default values for medium breasts.

Breast shape	Number	Location xc, yc, zc, mm	Dimensions xr, yr, zr, mm	Angles, in degrees, ϕ, θ, ψ	Att coeff, mm^{-1}	Colour
Breast semi-ellipsoid	1	20, 0, 0	70, 60, 60	−90°, 95°, 0°	0	Fuchsia
Breast hyperboloid	1	40, 0, 80, 11	7, 7, 5, 10	0°, 0°, 0°	0	Fuchsia
Breast nipple	1	−39, 0, −4	5, 5, 6	−90°, 95°, 0°	0	Fuchsia
Duct tree—major ducts	3	−39.4, 0, −4		−10°, −20°, 0°	0.037	Green

Table 5.8. Default values for large breasts.

Breast shape	Number	Location xc, yc, zc, mm	Dimensions xr, yr, zr, mm	Angles, in degrees, ϕ, θ, ψ	Att coeff, mm^{-1}	Colour
Breast semi-ellipsoid	1	50, 0, 0	70, 70, 100	−90°, 100°, 0°	0	Fuchsia
Breast hyperboloid	1	43, 0, 90	7, 7, 8, 8	0°, 0°, 0°	0	Fuchsia
Breast nipple	1	−48, 0, −15	5, 5, 6	−90°, 105°, 0°	0	Fuchsia
Duct tree—major ducts	3	−48, 0, −15		−10°, −10°, 0°	0.037	Green

5.5 Visualisation of computational breast models

There are several utilities that are used to visualise 3D breast models, visualising the CSG computational breast model (*.*phn* file) and the voxel breast phantom.

To load a CSG format breast phantom, which is designed as a combination of 3D geometrical primitives, the *Open Phantom* option from the main menu is used (figure 5.4(a)). Consequently, the phantom is visualised by using the *Visualization* menu (figure 4.12, chapter 4) from the *XRAYImagingSimulator*. Alternatively, the breast phantom may be visualised by using the *ImageJ* application.

5.6 Simulation of x-ray breast imaging

The simulation of the radiographic imaging process can be accomplished by choosing the *Mammography Simulation* option from the main menu (figure 5.17).

The *Mammography Simulation* panel consists of two sub-panels: one is dedicated to loading the 3D breast matrix (figure 5.18) and the second to generating x-ray projection images. Breast models are saved in a 3D matrix containing attenuation absorption coefficients in mm^{-1}.

When the breast voxel model is loaded successfully, the *Mammography Simulation* panel expands, as shown in figure 5.19. An additional panel *Irradiation Properties* appears dedicated to simulation of x-ray projection images from the loaded voxel breast model. This panel contains a set of edit and check boxes, used to define the parameters of the simulated imaging geometry. The panel was initially designed for isocentric imaging geometry, as depicted in the figure.

Figure 5.20 shows the simulated geometry of the imaging process for 3D mammography and that for isocentric imaging. SID and SDD are the distances from the source to the centre of rotation and to the detector plane, respectively. Monochromatic beams emerge from a point source, at a distance SID from the object plane, and penetrate the breast phantom. Photon transport is calculated as the transmitted intensity reaching the detector, based on exponential attenuation of monochromatic beams. In 3D imaging mode, angular projections may be obtained by rotating the source by the corresponding angle. The detector may remain still, as

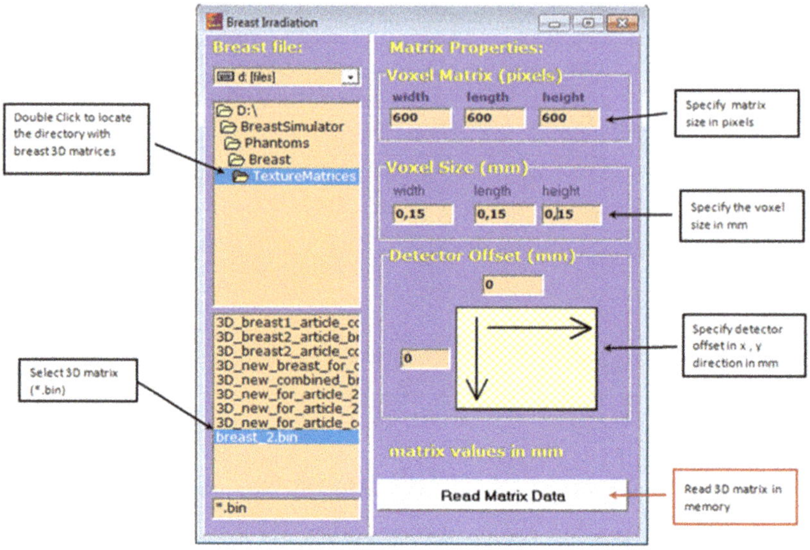

Figure 5.18. Loading the voxel breast attenuation matrix.

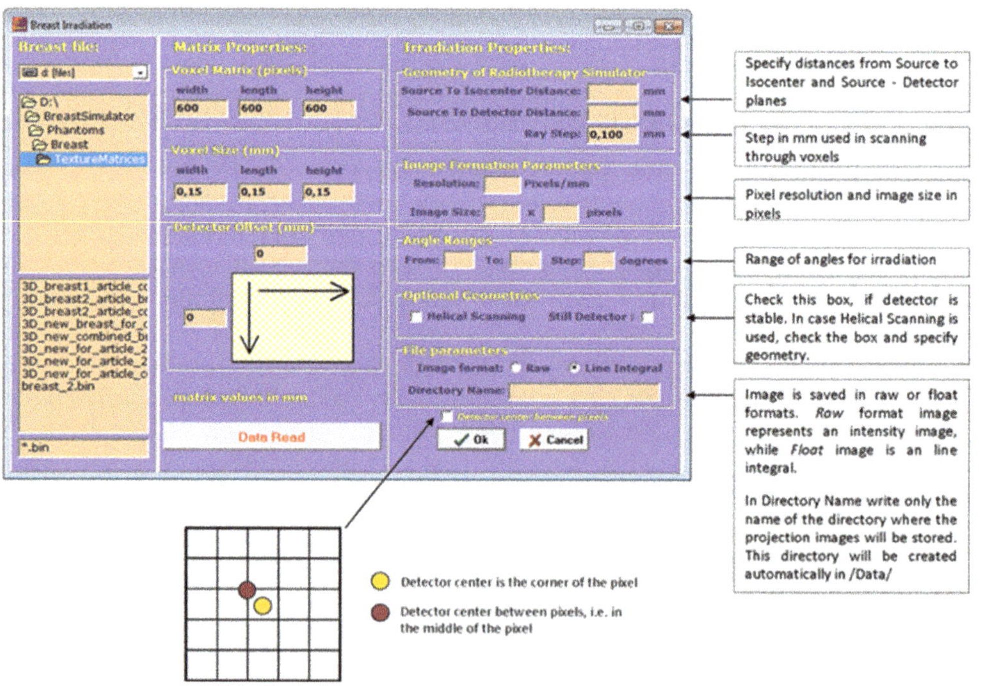

Figure 5.19. Full mammography image simulation interface.

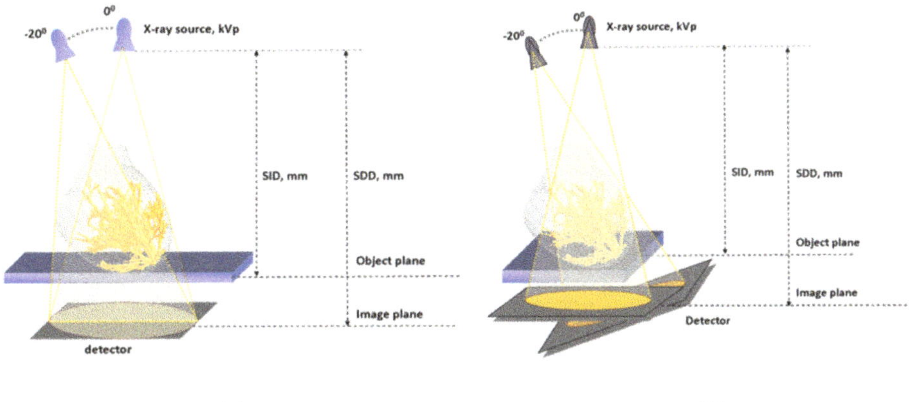

(a) (b)

Figure 5.20. Simulated irradiation geometry: (a) simulation of mammography imaging with a stable detector and (b) simulation of breast imaging with isocentric motion.

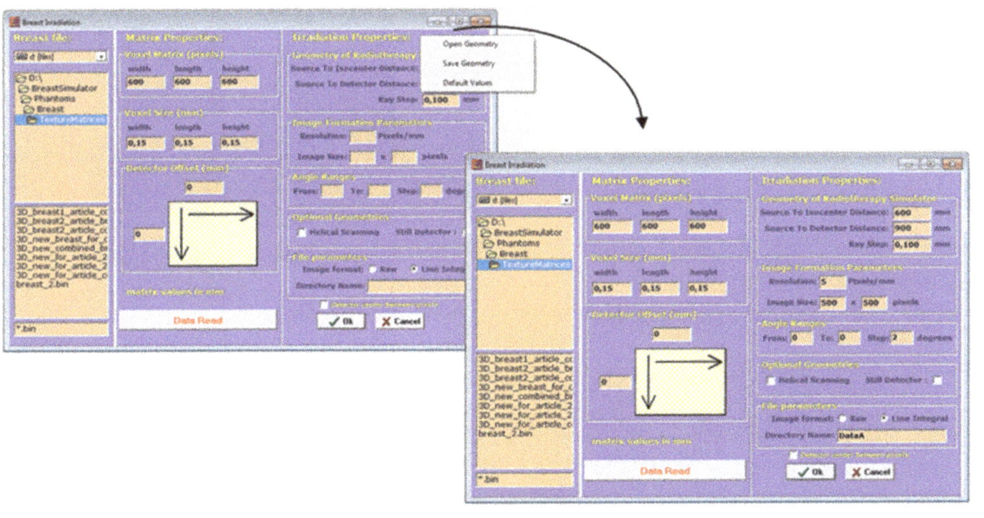

Figure 5.21. Loading default imaging parameters.

in the case of mammography, or rotate isocentrically coupled to the x-ray source. Scatter and detector responses are not simulated.

For *Default Geometry Parameters* which are for isocentric motion, a right-click of the mouse on the panel *Irradiation Properties* should be used (figure 5.21).

New imaging parameters are stored using the *Save Geometry* option. The file **.geo* is a binary file that contains the whole irradiation geometry structure and can be found in the */Geometry* directory as shown in figure 5.22. Figure 5.23 demonstrates the use of the *Save Geometry* option.

User-defined geometry can be uploaded using the *Open Geometry* option. In the case of helical imaging, the parameters are specified by using the check box *Helical Scanning* (figure 5.24).

The parameters used to describe and simulate helical scanning are graphically depicted in figure 5.24(b). In this figure, the spiral graphically depicts the helical

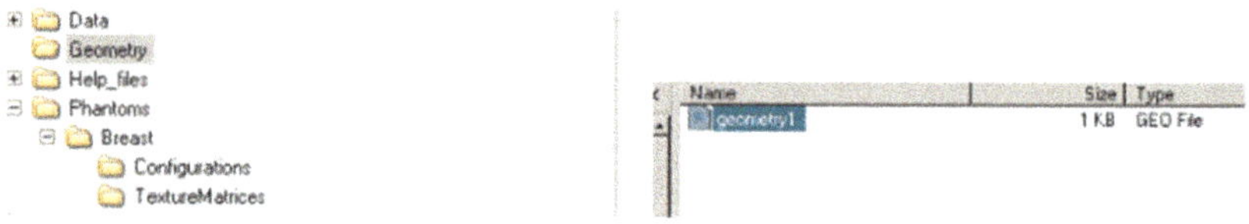

Figure 5.22. Irradiation geometry stored to the *geometry1.geo* file.

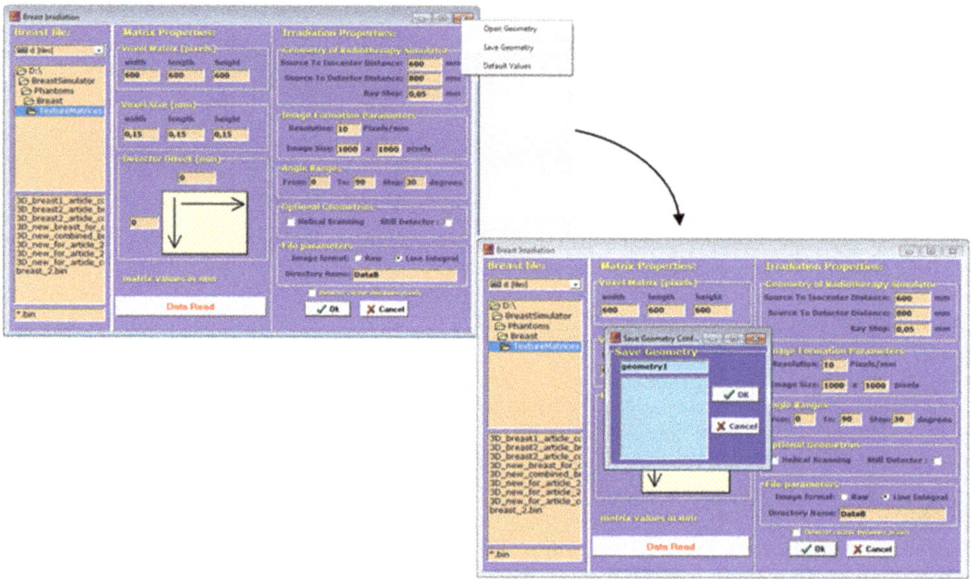

Figure 5.23. The use of *Save Geometry* options.

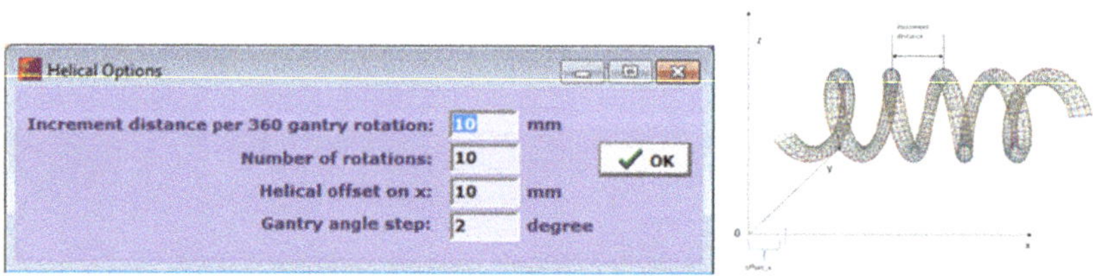

(a) (b)

Figure 5.24. Implementation of helical trajectory imaging: (a) parameters in the program and (b) a sketch of helical movement and parameters that correspond to parameters in the *Helical Geometry* options menu.

movement of the source–detector assembly. The movement is in the x-direction and starts from a position denoted by *Helical offset on x*. The number of full 360° rotations and the step between these rotations (increment distance) and the number of images per full 360° rotation (defined through the parameter gantry angle step) comprise the complete set of parameters needed to simulate images with helical scanning.

5.7 X-ray image visualisation

There are several possibilities to display images obtained from simulations: by using commercially available software such as *ImageJ, MATLAB*, or dedicated modules within the software, showing mammographic projection images, as well as the possibility of creating dual-energy images.

5.7.1 Single x-ray projection image

To display projection images, simulated with the *Irradiation* module, select the *Tools/Image Show* option from the main menu (figure 5.25). The simulated projection image (saved as 32 bit real) can be displayed in this module (figure 5.26).

The *Rescale* button invokes a simple image enhancement technique based on linear contrast stretching between values defined as *min_low* and *max_low*. As a result of this function, the displayed image is improved, as shown in figure 5.27.

This image can be further saved or copied by clicking the right mouse button when the mouse is on the image.

5.7.2 Dual-energy imaging

For dual-energy applications, there is a dedicated module to perform dual-energy imaging based on projections, acquired with two energies. In this case, a simple subtraction algorithm and weighting of the initial projections is applied. The *Tools/Dual Energy Imaging* option is used to compute and display the subtracted image from two projection images simulated or experimentally obtained (figure 5.28).

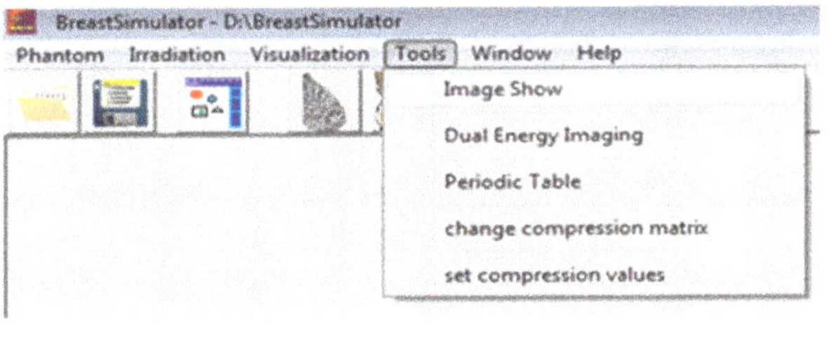

Figure 5.25. Image display of a single projection image.

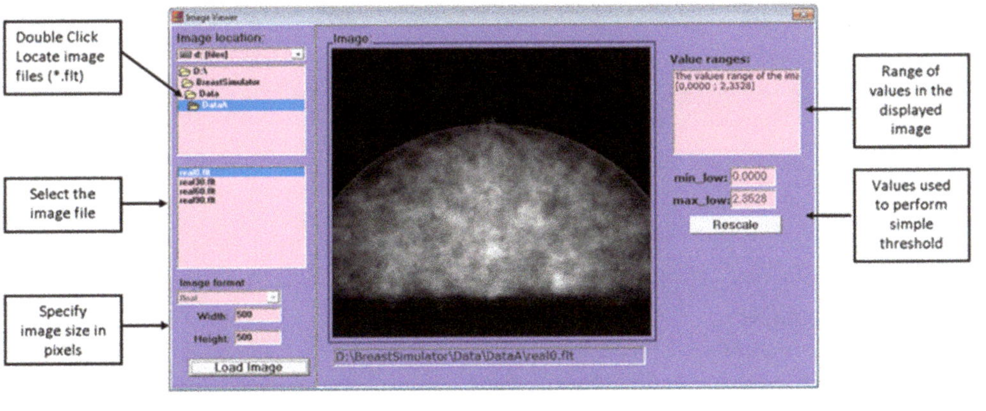

Figure 5.26. The *Image Show* panel, available for displaying projection images.

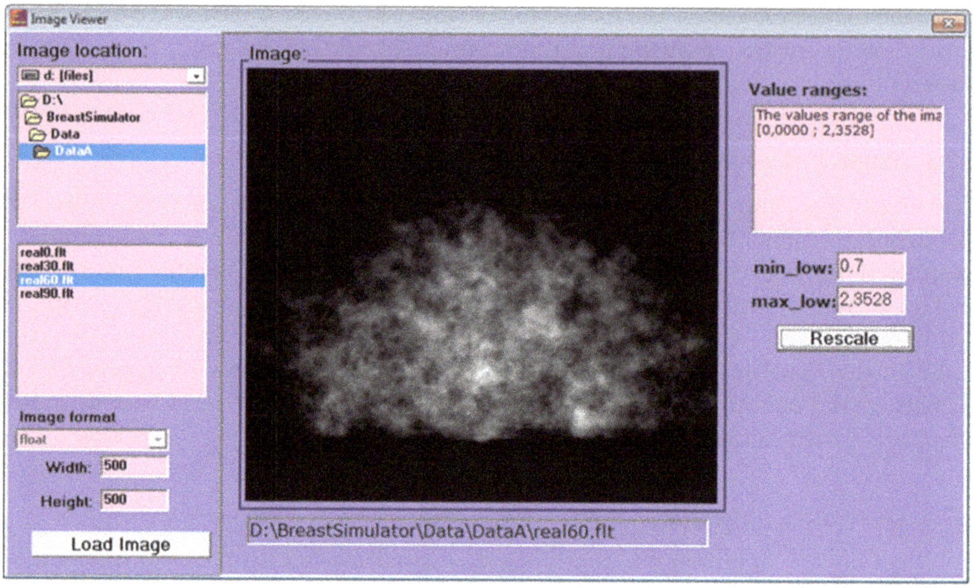

Figure 5.27. Image enhancement using the linear contrast stretching technique.

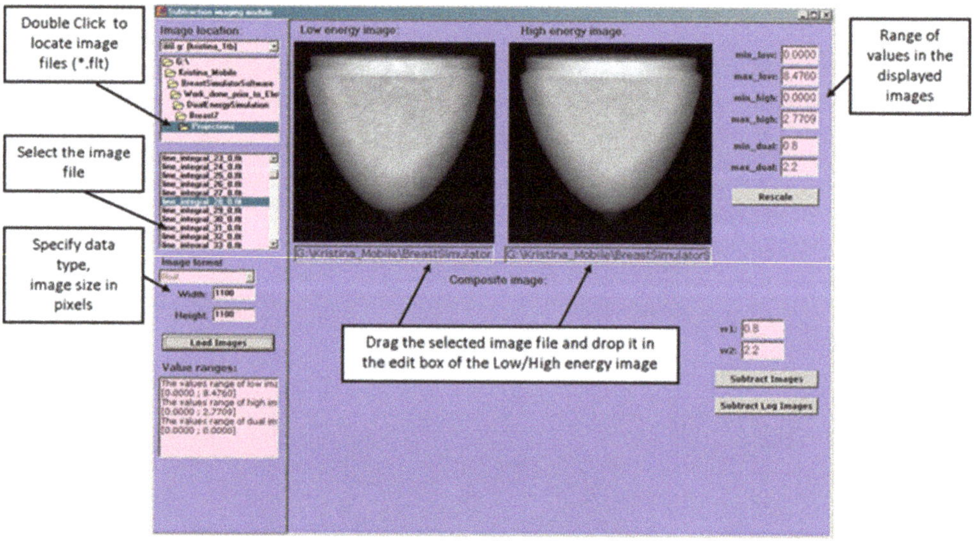

Figure 5.28. Dual-energy panel arrangement. The upper part of the panel contains the two images for subtraction.

Use *Load Images* to load the 'low' and 'high' images. Before that, they must be specified in the corresponding edit boxes. To compute the subtraction image two choices are available:

(a) *Subtract images*:

$$\text{Subtracted image} = w_1 I_{\text{L}} - w_2 I_{\text{H}}, \tag{5.6}$$

where I_L is the image obtained at low energy, while I_H is the image obtained at high energy.

(b) *Subtract log images:*

$$\text{Subtracted image} = w_2 \log(I_H) - w_1 \log(I_L). \tag{5.7}$$

These formulas perform the so-called linear subtraction algorithm (Boone and Shaber 1990). Case (a) is used when images are linear integrals, while case (b) is used when images are intensity images.

The coefficient w_2 is set to 1, while the coefficient w_1 is $\mu_{\text{breast}}(E_H)/\mu_{\text{breast}}(E_L)$. The result of the subtraction of the two images from the example in figure 5.28 is shown in figure 5.29. For this specific example the 'low' and the 'high' images are simulated for x-ray beams with energies of 17 keV and 28 keV, respectively. Their combination results in a dual-energy image with improved microcalcification visibility.

Similarly to the single image display, the right mouse on each of these images will invoke a pop-up menu with choices to save or copy the image to the clipboard. Additional information for particular case studies is discussed in chapter 7.

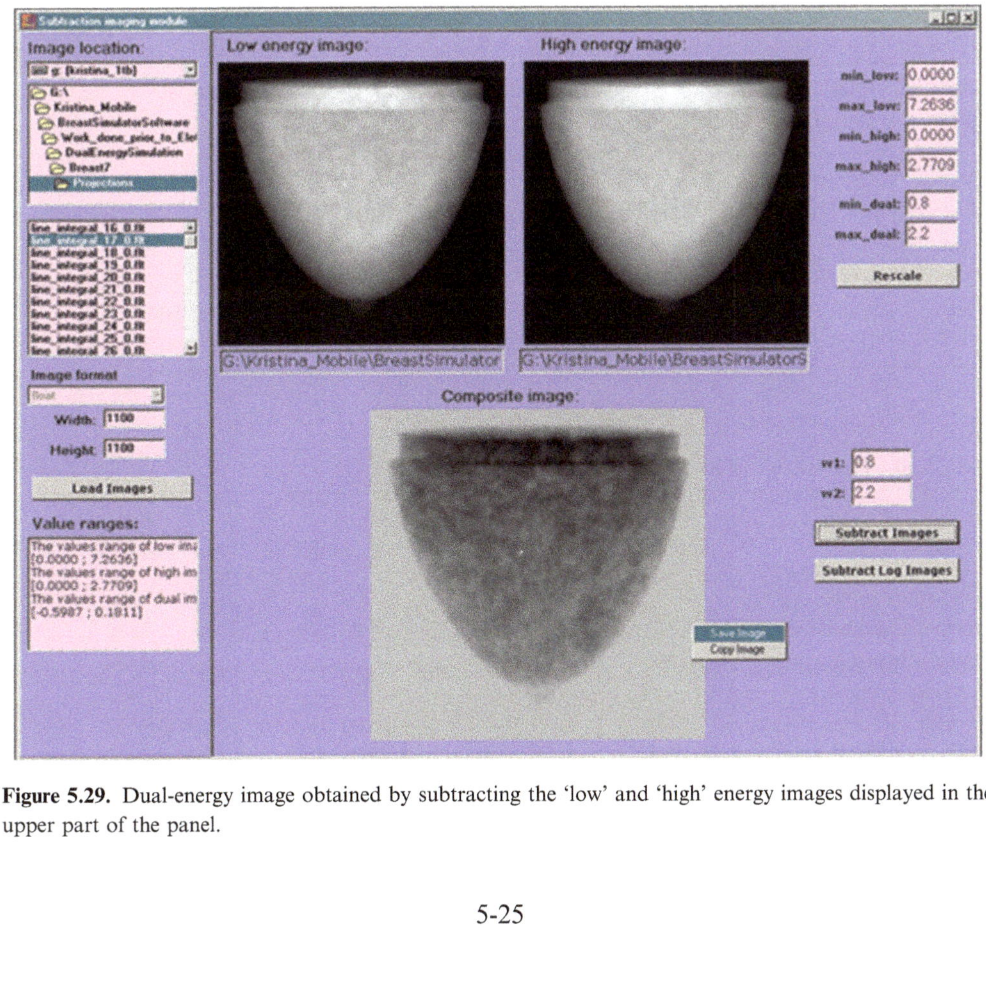

Figure 5.29. Dual-energy image obtained by subtracting the 'low' and 'high' energy images displayed in the upper part of the panel.

5.8 Examples and parameter estimation

Table 5.9 summarises the default values for the creation of various breasts with different densities and corresponding dimensions. They were obtained as a result of simulation studies, which aimed to set up the default values for small, medium, and large breast models and to generate mammographic images with realistic structure appearance. The creation of fatty, glandular, or dense breasts is influenced by the various combinations of the values of the following modelling parameters: the number of walks (N_{walks}), the number of steps per walk (N_{steps}), the Hurst exponent (H), the increment (h), the number of Cooper ligaments, and the duct density. The voxel dimension of the synthesised breast models varies from 0.1 to 0.15 mm, with the first value assigned to small and medium breasts, while the second value is used for large breast models. The breast dimensions correspond to that of the elongated semi-ellipsoid.

Examples of fatty, glandular, and dense breast models and simulated mammographic images are given in the following examples. Table 5.10 summarises the attenuation coefficients of the simulated tissues for 19 keV x-ray energy.

Table 5.9. Default values for the creation of breast models of different glandularity and dimensions. The following abbreviations are used: F—fatty breast; G—glandular breast; D—dense breast; matrix size is in thee dimensions.

Matrix size	N_{walk}	N_{steps}	h	H	Breast density	Breast size, mm	Breast type
700	1000	$(1–3) \times 10^3$	0.1	0.8	D	$40 \times 40 \times 40$	Small
1000	300–500	$(1–3) \times 10^4$	0.1, 0.2	0.8, 0.9	F		
1000	500–1000	$(1–2) \times 10^4$	0.1	0.9, 0.95	G		
1000	5000	2×10^4	0.1	0.9	D		
1200	500–1000	2×10^5	0.1	0.9	D	$50 \times 50 \times 80$	Medium/
1500	1000	$(1.5–2) \times 10^5$	0.1	0.9	G	$70 \times 70 \times 100$	large
1500	1000	$(1–1.5) \times 10^5$	0.1	0.9	F		

Table 5.10. Attenuation coefficients at 19 keV incident energy.

Tissue	Attenuation coefficient, $cm^2\ g^{-1}$	ρ, g cm^{-3}	Attenuation coefficient, cm^{-1}
Glandular tissue	0.772	1.02	0.7874
Cooper ligaments	0.772	1.02	0.7874
Pectoralis muscle	0.926	1.05	0.9723
Adipose tissue	0.630	0.95	0.5985
Lymph	0.935	1.03	0.9631
Blood	0.934	1.06	0.9900
Skin	0.732	1.09	0.7979

Table 5.11. Parameters for obtaining simulated images used in most of the applications.

Parameter	Default value
Source to object plane distance	600 mm
Source to detector plane distance	700 mm
Ray step	0.1 mm
Pixel resolution	[10–20] pixels/mm
Image size	1500 pixels

Table 5.11 lists the default parameters used in most of the image formation simulations.

5.8.1 Creation of a large fatty breast

The aim is to generate a large fatty breast with the parameters shown below. Cooper ligaments are not modelled.

Semi-elipsoid - part 1

(x0, y0, z0) = (50,0,0);

(xr, yr, zr) = (70,70,100);

(ϕ, θ) = (-90, 100);

att coeff = 0;

Semi-hyperboloid - part 2

(x0,y0,z0,free_offset) = (43,0, 90, 20);

(xr, yr, zr, c_param) = (7, 7 ,5, 8);

(ϕ, θ) = (0, 0, 0);

att coeff = 0;

Nipple

(x0, y0, z0) = (-48, 0, -15);

(xr, yr, zr) = (5, 4, 4);

(ϕ, θ) = (0, -90, 105);

att coeff = 0;

Ducts

(x0, y0, z0) = (-48, 0, -15); (ϕ, θ) = (-10, -10); att_coeff = 0.06 mm^{-1}

number of major ducts	height of major ducts, mm	radius of major ducts, mm	number of lact. ducts	initial height, mm	limit height, mm	duct initial radius, mm	lobule, radius, mm
5	1	1.6	6	12	2	0.15	0.4

Texture

att_coeff = 0.002 – 0.05 mm^{-1}

size	voxel, mm	N_{walk}	N_{step}	window	increment	Hurst coeff	StDev
1500^3	0.15	10^3	$200x10^3$	3	0.1	0.9	1.5

The breast model generated with these parameters is shown in figure 5.30(a), while its x-ray projection image, which corresponds to a craniocaudal view, is shown in figure 5.30(b).

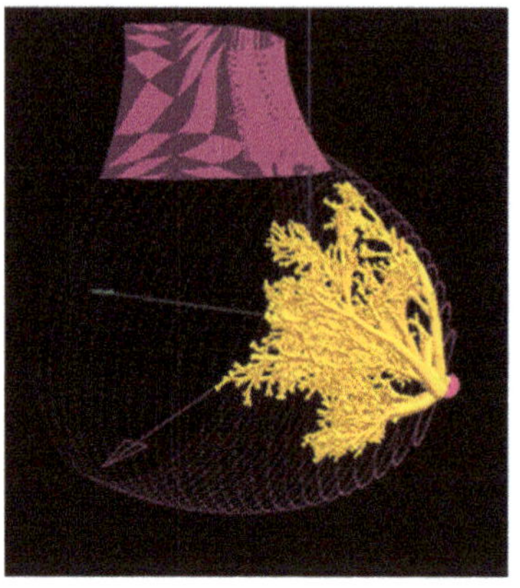

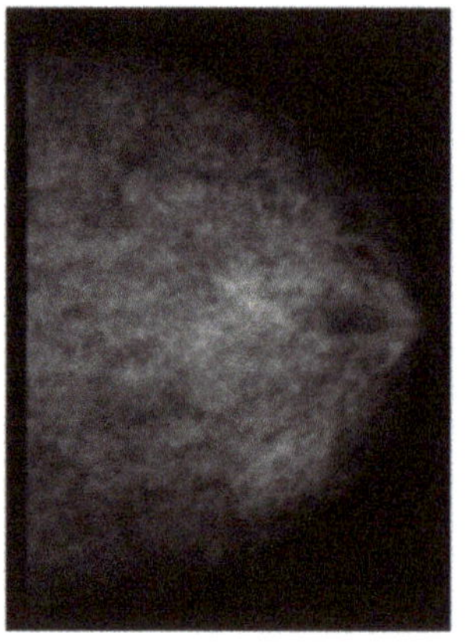

(a) (b)

Figure 5.30. Generated large fatty breast: (a) computational breast model and (b) x-ray projection images.

5.8.2 Creation of a small dense breast

The aim is to generate a small dense breast with the parameters shown below.

Semi-ellipsoid - part 1

(x0, y0, z0) = (24,0,0);

(xr, yr, zr) = (40,40,40);

(ϕ, θ) = (-90, 90);

att coeff = 0;

Semi-hyperboloid - part 2

(x0,y0,z0,free_offset) = (25,0, 55, 11);

(xr, yr, zr, c_param) = (5, 5 ,5, 6);

(ϕ, θ) = (0, 0, 0);

att coeff = 0;

Nipple

(x0, y0, z0) = (-15, 0, -2);

(xr, yr, zr) = (4, 4, 5);

(ϕ, θ) = (0, -90, 90);

att coeff = 0;

Ducts

(x0, y0, z0) = (-14.5, 0, -2); (ϕ, θ) = (-10, -10); att_coeff = 0.06 mm^{-1}

number of major ducts	height of major ducts, mm	radius of major ducts, mm	number of lact. ducts	initial height, mm	limit height, mm	duct initial radius, mm	lobule, radius, mm
6	1	2	6	6	2.2	0.12	0.5

Cooper ligaments

number = 25000 ; radius = 10 mm ; att_coeff = 0.06 mm^{-1}

Texture

att_coeff = 0.002 – 0.05 mm^{-1}

size	voxel, mm	N_{walk}	N_{step}	low pass	dilation	window	increment	Hurst coeff	StDev
600^3	0.15	2000	1200	yes	yes	4	0.15	0.8	1.5

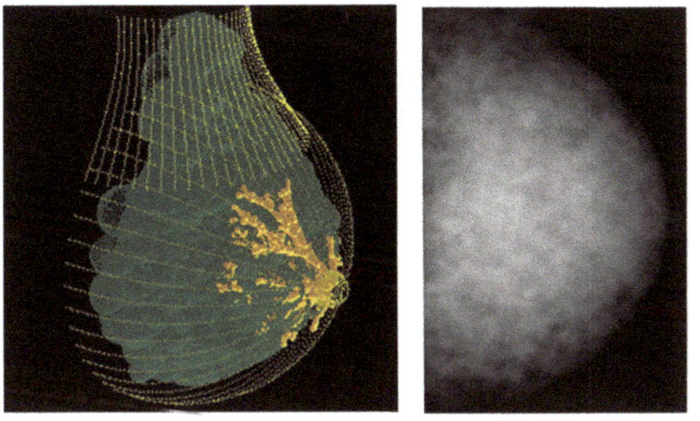

(a) (b)

Figure 5.31. Generated small dense breast: (a) breast model with no texture (*.*phn*) and (b) a mammography projection image.

A breast model generated with these parameters is shown in figure 5.31(a), while its x-ray projection image, which corresponds to a craniocaudal view, is shown in figure 5.31(b).

5.8.3 Creation of a small glandular breast

The aim is to generate small glandular breast with the parameters shown below.

Semi-ellipsoid - part 1	Semi-hyperboloid - part 2	Nipple
(x0, y0, z0) = (24,0,0);	(x0,y0,z0,free_offset) = (25,0, 55, 11);	(x0, y0, z0) = (-15, 0, -2);
(xr, yr, zr) = (40,40,40);	(xr, yr, zr, c_param) = (5, 5 ,5, 6);	(xr, yr, zr) = (4, 4, 5);
(ϕ, θ) = (-90, 90);	(ϕ, θ) = (0, 0, 0);	(ϕ, θ) = (0, -90, 90);
att coeff = 0;	att coeff = 0;	att coeff = 0;

Ducts

(x0, y0, z0) = (-14.5, 0, -2); (ϕ, θ) = (-10, -10); att_coeff = 0.06 mm^{-1}

number of major ducts	height of major ducts, mm	radius of major ducts, mm	number of lact. ducts	initial height, mm	limit height, mm	duct initial radius, mm	lobule, radius, mm
5	1	2	5	6	2.2	0.12	0.5

Cooper ligaments

number = 50000 ; radius = 3 mm ; att_coeff = 0.06 mm^{-1}

Texture

att_coeff = 0.002 – 0.05 mm^{-1}

size	voxel, mm	N_{walk}	N_{step}	low pass	dilation	window	increment	Hurst coeff	StDev
600^3	0.15	800	1000	yes	yes	4	0.15	0.8	1

The breast model generated with these parameters is shown in figures 5.32(a), (b), while its x-ray projection image, which corresponds to a lateral view is shown in figure 5.32(c).

5.8.4 Creation of a small glandular breast with lymphatics

The aim is to generate small glandular breast with a lymphatic system and the parameters shown below.

Semi-ellipsoid - part 1	Semi-hyperboloid - part 2	Nipple
(x0, y0, z0) = (14,0,0);	(x0,y0,z0,free_offset) = (15,0, 55, 11);	(x0, y0, z0) = (-25, 0, -2);
(xr, yr, zr) = (40,40,40);	(xr, yr, zr, c_param) = (5, 5 ,5, 6);	(xr, yr, zr) = (2.5, 3, 3);
(ϕ, θ) = (-90, 90);	(ϕ, θ) = (0, 0, 0);	(ϕ, θ) = (0, -90, 90);
att coeff = 0;	att coeff = 0;	att coeff = 0;

Ducts

(x0, y0, z0) = (-25, 0, -2); (ϕ, θ) = (-10, -10); att_coeff = 0.06 mm^{-1}

number of major ducts	height of major ducts, mm	radius of major ducts, mm	number of lact. ducts	initial height, mm	limit height, mm	duct initial radius, mm	lobule, radius, mm
4	1	1.3	5	5.5	1.5	0.14	0.3

Cooper ligaments

number = 40000 ; radius = 3 mm ; att coeff = 0.06 mm^{-1}

Lymphatics

number = 3 ; radius = 0.15 mm ; initial_height = 6 mm; end_height = 2.0 mm; att_coeff = 0.1 mm^{-1}

Texture

att_coeff = 0.02 – 0.07 mm^{-1}

size	voxel, mm	N_{walk}	N_{step}	low pass	dilation	window	increment	Hurst coeff	StDev
600^3	0.15	800	800	yes	yes	2	0.1	0.99	0.0

The breast generated with these parameters is shown in figure 5.33(a, b), while its x-ray projection image, which corresponds to a lateral view, is shown in figure 5.33(c).

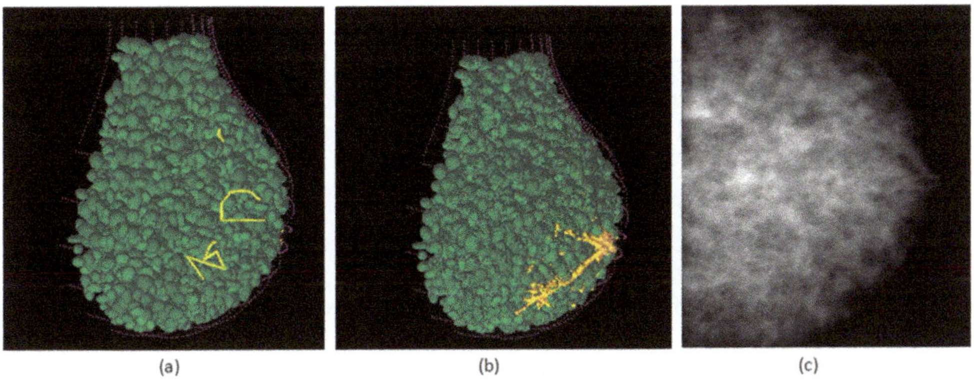

Figure 5.32. Generated small dense breast: (a,b) breast model with no texture (*.*phn*) and (c) mammography projection image.

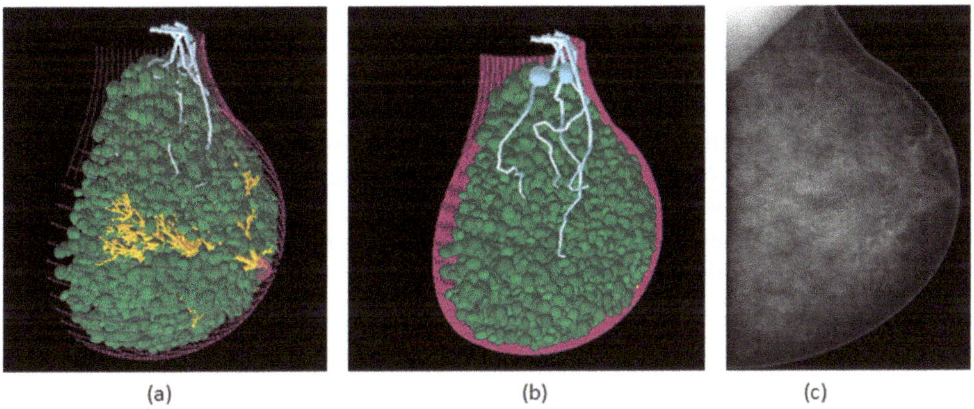

Figure 5.33. Generated small dense breast: (a, b) breast model with no texture (*.*phn*) and (c) mammography projection image.

5.9 Self-training exercises

Exercise 1: Creation of a breast model

Self-study video tutorials are available at the following links:

http://tuv.eutempe-net.eu/wp-content/uploads/Education-materials/Videos/ComposeBreast%20reduced.mp4.

http://tuv.eutempe-net.eu/wp-content/uploads/Education-materials/Videos/BreastVisualization%20optimized.mp4.

The aim of this self-training exercise is to create a small breast phantom.

Tasks:

(a) Run the *BreastSimulator*.
(b) From the application menu, select *Phantom* → *Compose Breast* option.
(c) From the application menu, select *Breast Configuration* → *Open default small breast*.
(d) Keep the default values.

(e) In the *Duct System* panel use the following parameters: number of major ducts = 3; number of lactiferous ducts = 3; max height = 6; limit height = 2.2.
(f) Generate the breast model by using the *Generate Breast Model* button.
(g) Open the created **.phn* phantom by choosing the *Open phantom* option. Select *breast_3D_new.phn.*
(h) Visualise the generated model.

Exercise 2: Generation of a mammography image

The aim of this self-training exercise is to generate an x-ray image of a voxel breast phantom.

Self-study video tutorials are available at the following link:
http://tuv.eutempe-net.eu/wp-content/uploads/Education-materials/Videos/ImageFormation%20optimized.mp4.

Tasks:

(a) Run the *BreastSimulator.*
(b) From the application menu, select the *Analytical Image Simulation* → *Image Simulation* option.
(c) Set the directory *.../BreastSimulator/Phantoms/Breast/TextureMatrices* and load *breast_1.bin.* The voxel matrix has dimensions of: 600 × 600 × 600 voxels; voxel size: 0.15 mm in all directions.
(d) For the geometry parameters, use the same values as those in figure 5.18:
  - SID = 600 mm; SDD = 900 mm.
  - Ray step = 0.1 mm.
  - Resolution = 5 pixel/mm.
  - Image size = 500 × 500 pixels.
(e) Generate synthetic images.
(f) Visualise the results by using *Tools* → *Image Show.*

Exercise 3: Creation of a breast phantom

The aim of this self-training exercise is to create a small breast phantom.

Tasks:

(a) Run the *BreastSimulator.*
(b) From the application menu, select the *Phantom* → *Compose Breast* option.
(c) Parameters for the model:
  - Energy of the x-rays: 19 keV.
  - Semi-ellipsoid: ($x0$, $y0$, $z0$) = (24, 0, 0); (xr, yr, zr) = (40,40,40).
  - Semi-hyperboloid: ($x0$, $y0$, $z0$, *free_offset*) = (25, 0, 55, 11); (xr, yr, zr, *c_param*) = (5, 5, 5, 6).
  - Nipple: ($x0$, $y0$, $z0$) = (−15, 0, −2); (xr, yr, zr) = (4, 4, 5).
  - Cooper ligaments: number = 0; radius = 3 mm; att_coeff = 0.079 mm^{-1}.
  - Pectoralis: no.
  - Skin: no.
  - Ducts: ($x0$, $y0$, $z0$) = (−14.5, 0, −2); (φ, θ) = (−10°, −10°); att_coeff = 0.079 mm^{-1}.

Major ducts	Major ducts, height [mm]	Major ducts, radius [mm]	Lact. Ducts	Lact. Ducts, in. height [mm]	Lact. Ducts lim. height [mm]	Lact. Ducts in. radius [mm]	Lobule radius [mm]
4	1	2	4	6	2.3	0.12	0.5

- Texture: size = 500 × 500 × 500 pixels; att_coeff = 0.02–0.079 mm^{-1}; matrix name: *small_breast_exercise3.*

Voxel [mm]	N_{walk}	N_{step}	LPF	Dilation	Filt. Size	Increment	Hurst coeff.	StDev
0.20	10	10	No	No	0	0.15	0.8	0

(d) Save the configuration as *exercise3.bin.*
(e) Generate the breast model using the *Generate Breast Model* button.
(f) Open the created **.phn* phantom by choosing the *Open phantom* option. Select *small_breast_exercise3.phn.*
(g) Visualise the generated model.

Exercise 4: Creation of breast phantom

The aim of this self-training exercise is to create a small dense breast phantom. For this purpose, open the already saved configuration *exercise3.bin.*

Tasks:

(a) Run the *BreastSimulator.*
(b) From the application menu, select *Phantom → Compose Breast* option.
(c) Parameters for the model:
   - Energy of the x-rays: 19 keV.
   - Semi-ellipsoid: ($x0$, $y0$, $z0$) = (24, 0, 0); (xr, yr, zr) = (40, 40, 40).
   - Semi-hyperboloid: ($x0$, $y0$, $z0$, *free_offset*) = (25, 0, 55, 11); (xr, yr, zr, *c_param*) = (5, 5, 5, 6).
   - Nipple: ($x0$, $y0$, $z0$) = (−15, 0, −2); (xr, yr, zr) = (4, 4, 5).
   - Cooper ligaments: number = 50 000; radius = 2 mm; att_coeff = 0.079 mm^{-1}.
   - Pectoralis: yes; att_coeff = 0.097 mm^{-1}.
   - Skin: yes; att_coeff = 0.08 mm^{-1}.
   - Ducts: ($x0$, $y0$, $z0$) = (−14.5, 0, −2); (φ, θ) = (−10°, −10°); att_coeff = 0.079 mm^{-1}.

Major ducts	Major ducts, height [mm]	Major ducts, radius [mm]	Lact. Ducts	Lact. Ducts, in. height [mm]	Lact. Ducts lim. height [mm]	Lact. Ducts in. radius [mm]	Lobule radius [mm]
4	1	2	4	6	2.6	0.12	0.5

- Texture: size = 500 × 500 × 500 pixels; att_coeff = 0.04–0.079 mm^{-1}; matrix name: *small_breast_exercise4.*

Voxel [mm]	N_{walk}	N_{step}	LPF	Dilation	Filt. Size	Increment	Hurst coeff.	StDev
0.20	1000	1000	yes	yes	3	0.1	0.9	1.5

(d) Generate the breast model by using *Generate Breast Model* button.
(e) Save the configuration file by selecting *Breast Configuration* → *Save user defined configuration* and choose a name for your configuration file (for example *exercise4*).
(f) Open the created **.phn* phantom by choosing the *Open phantom* option. Select *small_breast_exercise4.phn.*
(g) Visualise the generated model.

Exercise 5: Creation of a breast phantom

The aim of this self-training exercise is to create a small dense breast phantom with an abnormality. For this purpose, open the already saved configuration *exercise3.bin.*

Tasks:

(a) Run the *BreastSimulator.*
(b) From the application menu, select the *Phantom* → *Compose Breast* option.
(c) Parameters for the model:
- Energy of the x-rays: 19 keV.
- Semi-ellipsoid: ($x0$, $y0$, $z0$) = (24, 0, 0); (xr, yr, zr) = (40, 40, 40).
- Semi-hyperboloid: ($x0$, $y0$, $z0$, free_offset) = (25, 0, 55, 11); (xr, yr, zr, c_param) = (5, 5, 5, 6).
- Nipple: ($x0$, $y0$, $z0$) = (−15, 0, −2); (xr, yr, zr) = (4, 4, 5).
- Cooper ligaments: n = 25 000; radius = 3 mm; att_coeff = 0.079 mm^{-1}.
- Pectoralis: no.
- Skin: no.
- Ducts: ($x0$, $y0$, $z0$) = (−14.5, 0, −2); (φ, θ) = (−10°, −10°); att_coeff = 0.079 mm^{-1}.

Major ducts	Major ducts, height [mm]	Major ducts, radius [mm]	Lact. Ducts	Lact. Ducts, in. height [mm]	Lact. Ducts lim. height [mm]	Lact. Ducts in. radius [mm]	Lobule radius [mm]
4	1	2	4	6	2.2	0.12	0.5

- Texture: size = 500 × 500 × 500 pixels; att_coeff = 0.04–0.079 mm^{-1}; matrix name: *small_breast_exercise5.*

Voxel [mm]	N_{walk}	N_{step}	LPF	Dilation	Filt. Size	Increment	Hurst coeff.	StDev
0.15	500	1000	yes	yes	4	0.1	0.8	1.5

   - Abnormalities: clusters = 1; cluster radius = 10 mm; μCalc/Cluster = 10; μCalc radius = [0.1 : 0.8] mm; att_coeff = 1.845 mm^{-1}.

(d) Generate the breast model using the *Generate Breast Model* button.
(e) Save the configuration file by selecting *Breast Configuration* → *Save user defined configuration* and choose a name for your configuration file (for example *exercise5*).
(f) Open the created **.phn* phantom by choosing the *Open phantom* option. Select *small_breast_example5.phn.*
(g) Visualise the generated model.

Exercise 6: Creation of a breast phantom

The aim of this self-training exercise is to create a large fatty breast phantom. For this purpose, open the already saved configuration *exercise3.bin.*

Tasks:

(a) Run the *BreastSimulator.*
(b) From the application menu, select the *Phantom* → *Compose Breast* option.
(c) Parameters for the model:
   - Energy of the x-rays: 19 keV.
   - Semi-ellipsoid: ($x0$, $y0$, $z0$) = (50, 0, 0); (xr, yr, zr) = (70, 70, 100).
   - Semi-hyperboloid: ($x0$, $y0$, $z0$, free_offset) = (43, 0, 90, 20); (xr, yr, zr, c_param) = (7, 7, 8, 8).
   - Nipple: ($x0$, $y0$, $z0$) = (−48, 0, −15); (xr, yr, zr) = (5, 5, 6).
   - Cooper ligaments: number = 0; radius = 3 mm; att_coeff = 0.079 mm^{-1}.
   - Pectoralis: no.
   - Skin: no.
   - Ducts: ($x0$, $y0$, $z0$) = (−48, 0, −15); (φ, θ) = (−10°, −10°); att_coeff = 0.079 mm^{-1}.

Major ducts	Major ducts, height [mm]	Major ducts, radius [mm]	Lact. Ducts	Lact. Ducts, in. height [mm]	Lact. Ducts lim. height [mm]	Lact. Ducts in. radius [mm]	Lobule radius [mm]
6	1	2	6	14	5	0.12	0.5

   - Texture: size = 256 × 256 × 256 pixels; att_coeff = 0.04–0.079 mm^{-1}, matrix name: *large_breast_exercise6.*

Voxel [mm]	N_{walk}	N_{step}	LPF	Dilation	Filt. Size	Increment	Hurst coeff.	StDev
1	100	1000	Yes	Yes	5	1	0.5	1

(d) Generate the breast model using the *Generate Breast Model* button.

(e) Save the configuration file by selecting *Breast Configuration* → *Save user defined configuration* and choose a name for your configuration file (for example *exercise6*).
(f) Open the created **.phn* phantom by choosing the *Open phantom* option. Select *large_breast_exercise6.phn.*
(g) Visualise the generated model.

Exercise 7: Creation of x-ray images

The aim of this self-training exercise is to generate an x-ray image of a voxel breast phantom.

Tasks:

(a) Run the *BreastSimulator*.
(b) From the application menu, select the *Analytical Image Simulation* → *Image Simulation* option.
(c) Use the *small_breast_example4.bin* phantom created in *exercise 4* and generate synthetic x-ray images of the breast phantom at 0° and 90° gantry angles for circular isocentric geometry, using the following settings:
   - SID = 600 mm; SDD = 900 mm; ray tracing step = 0.1 mm; image size: 500 × 500 pixels; resolution: 5 pixels/mm. Save the image in a folder *Excersise5_7a.*
   - SID = 600 mm; SDD = 900 mm; ray tracing step = 0.1 mm; image size: 1000 × 1000 pixels; resolution: 10 pixels/mm. Save the image in a folder *Excersise5_7b.*
   - SID = 600 mm; SDD = 900 mm; ray tracing step = 0.1 mm; image size: 2000 × 2000 pixels; resolution: 20 pixels/mm. Save the image in a folder *Excersise5_7c.*
(d) Compare the images from the different scenarios.

References

Bliznakova K *et al* 2003 A three-dimensional breast software phantom for mammography simulation *Phys. Med. Biol.* **48** 3699–3719

Bliznakova K *et al* 2012 BreastSimulator: a software platform for breast x-ray imaging research *J. Biomed. Graphics Comput.* **2** 1–14

Bliznakova K *et al* 2010 Evaluation of an improved algorithm for producing realistic 3D breast software phantoms: application for mammography *Med. Phys.* **37** 5604–5617

Boone J M and Shaber G S 1990 Dual-energy mammography: a detector analysis *Med. Phys.* **17** 665–675

De Paredes E S 1992 *Atlas of Film—Screen Mammography* (Baltimore, MD: Williams and Wilkins)

Egan R L 1988 *Breast Imaging: Diagnosis and Morphology of Breast Diseases* (Philadelphia, PA: W. B. Saunders)

Johns H E and Cunningham J R 1983 *Physics of Radiology* (Springfield, IL: Charles C Thomas)

Junqueira L C, Carneiro J and Kelley R O 1989 *Basic Histology* 6th edn (Norwalk, CT: Appleton & Lange)

Novak R 1988 Transformation of the female breast during compression at mammography with special reference to the importance for localization of a lesion *Acta Radiol. Suppl.* **371** 1–47

O'Connell A M *et al* 2014 The potential role of dedicated 3D breast CT as a diagnostic tool: review and early clinical examples *Breast J.* **20** 592–605

Olsen A H *et al* 2005 Breast cancer mortality in Copenhagen after introduction of mammography screening: cohort study *Bri. Med. J.* **330** 220

Schulz-Wendtland R *et al* 2009 Digital mammography: an update *Eur. J. Radiol.* **72** 258–265

Sechopoulos I 2013 A review of breast tomosynthesis. Part I. The image acquisition process *Med. Phys.* **40** 014301

Sickles E A 1986 Breast calcifications: mammographic evaluation *Radiology* **160** 289–93

Zonderland H and Smithuis R 2013 Bi-RADS for Mammography and Ultrasound 2013 Radiology Assistant

Chapter 6

Image reconstruction techniques

6.1 Introduction

The aim of this chapter is to help the reader understand the general idea of x-ray tomographic image reconstruction, the basic conditions and requirements, and factors that primarily influence the quality of the obtained images. A self-training example and an exercise will allow the reader to experience running relevant virtual studies.

6.2 X-ray computed tomography

6.2.1 Measurements and goals

Tomographic techniques provide images, representing values of different properties along planes of different orientation cutting the objects. To account for the 3D nature of the analysed objects and the limited resolution of the reconstructions, the tomographic images are also called slices.

Unlike, for example, conventional radiography, where the images are obtained directly, in a modern computed tomography modality the images are *computed* (obtained indirectly) using large sets of measured data.

Various object material properties can be evaluated and represented as images. The different tomographic imaging modalities perform measurements and compute and visualise different object or tissue properties (figure 6.1). Ultrasound imaging, MRI, SPECT, and PET are all tomographic techniques.

However, historically and traditionally, the term computed tomography (CT) has been related to the tomographic techniques that exploit the interactions of x-rays passing through matter.

doi:10.1088/2053-2563/aae197ch6

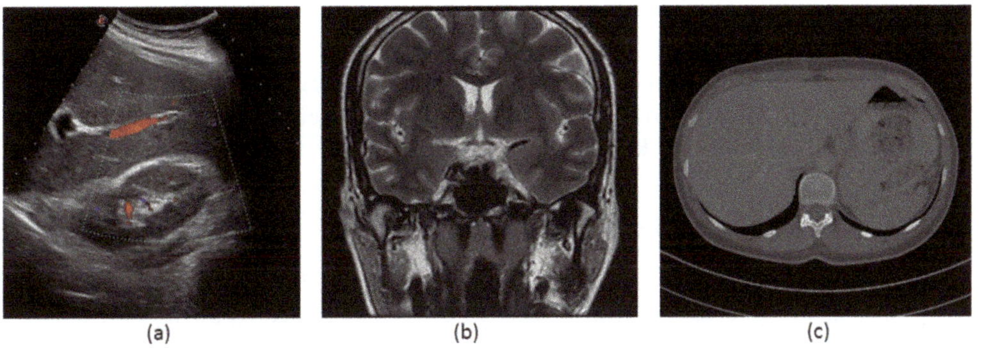

Figure 6.1. Examples of tomographic images from different modalities: (a) ultrasound, (b) MRI, and (c) x-ray CT.

6.2.2 Linear attenuation coefficient

The property of the objects/tissues that is computed and visualised in CT is the linear attenuation coefficient, μ, usually measured in [cm^{-1}] or [mm^{-1}]. This property is related to the percentage of interacting (scattered or absorbed) x-ray photons per unit path length when passing through a given material. For homogeneous materials this property remains constant, but in the general case (for complex objects with unknown internal structure, as well as for the human body) it varies and is normally considered as a three-dimensional function—$\mu(x, y, z)$. The values of μ also depend on the energy of the x-rays, but for the sake of presenting the basic principles and approaches, the use of monochromatic x-rays is considered next.

Obtaining a CT image is related to computing the values of μ within a preliminary selected plane in the space, passing through the object.

6.2.3 Hounsfield units

The linear attenuation coefficient of the soft tissues in the human body is comparable to that of water. Their relative difference is now widely used in CT (Bushberg 2002):

$$\mathrm{HU}(x, y, z) = 1000\frac{[\mu(x, y, z) - \mu_{\mathrm{w}}]}{\mu_{\mathrm{w}} - \mu_{\mathrm{a}}}, \tag{6.1}$$

where $\mu(x, y, z)$ is the linear attenuation coefficient of the object material or the tissue, μ_{w} is the linear attenuation coefficient of water, and μ_{a} is the is the linear attenuation coefficient of air. The new quantity has been named the *Hounsfield unit* (HU) in honour of Sir Godfrey Hounsfield, one of the inventors of x-ray computed tomography.

6.2.4 Scanning in CT

X-ray CT uses measurements of the intensity of the x-rays leaving the objects or the human body. Series of such x-ray images are needed and are obtained all around the object by moving either the source–detector pair or the object in a known way, called the *scanning trajectory*.

In the case of using monochromatic x-rays of energy ε, the intensity images acquired by the detector are related to the linear attenuation coefficient by the Lambert–Beer law (Macovski 1983):

$$I(x_d, y_d, z_d) = I_0(x_d, y_d, z_d)e^{-\int_{(x_s,y_s,z_s)}^{(x_d,y_d,z_d)} \mu(x,\, y,\, z)dxdydz}, \tag{6.2}$$

where $I(x_d, y_d, z_d)$ is the intensity registered by a detector pixel at location (x_d, y_d, z_d) in the presence of an object, $I_0(x_d, y_d, z_d)$ is the intensity that is/would be registered by the same detector pixel without the presence of an object between the detector and the x-ray source, located at (x_s, y_s, z_s). The integration is performed along the line connecting the locations of the source and the detector pixel.

By applying a logarithm and reworking, an expression for the integration along a ray linear attenuation coefficient can be derived:

$$\int_{(x_s,\, y_s,\, z_s)}^{(x_d,\, y_d,\, z_d)} \mu(x,\, y,\, z)dxdydz = \ln\left[\frac{I_0(x_d,\, y_d,\, z_d)}{I(x_d,\, y_d,\, z_d)}\right]. \tag{6.3}$$

The values are known as *line integrals* (LI) of μ. In the above expressions the linear attenuation coefficient of air is neglected.

Intensity images captured by the detectors are known as projection images. The LI images, obtained from them are also called projection images. CT scanning allows acquiring of LI projections all around the object.

6.2.5 The Radon transform

Scanning in CT and the subsequent image reconstructions are formalised in an abstract way in mathematics by the Radon transform. The general form of the Radon transform requires integration of function values along hyperplanes (Natterer 1986). When working with functions of two variables, i.e. two-dimensional space, the hyperplanes are lines and integrations are performed along those lines. Each line can be described by the coordinates of its distance vector (collinear to its normal vector, with a length equal to the distance of the line to the origin of the coordinate system). Thus, the RT of a function in the *object space*, e.g. $\mu(x, y)$, is described by a two-dimensional function $R_\mu(x, y)$ in the *Radon space*:

$$R_\mu(x, y) = \Re\{\mu(x, y)\}. \tag{6.4}$$

All lines that pass via point p in the object space are represented by points in the Radon space laying on a circumference with a diameter defined by the origin of the coordinate systems and the coordinates of the point $\mu(x_p, y_p)$ (figure 6.2).

The polar version $R_\mu(\theta, t)$ of $R_\mu(x, y)$ is known as a *sinogram* of $\mu(x, y)$. Figure 6.3 gives an example of an image only containing two pixels with nonzero values, as well as its Radon transform in polar and Cartesian coordinates.

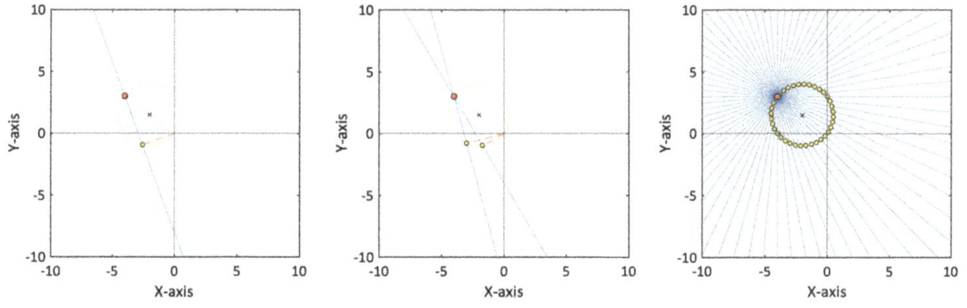

Figure 6.2. A point (in red) in the object space and a line (a), two lines (b) and more lines (c) through it. The lines are represented in the Radon space by the line points (in yellow) closest to the origin of the coordinate system.

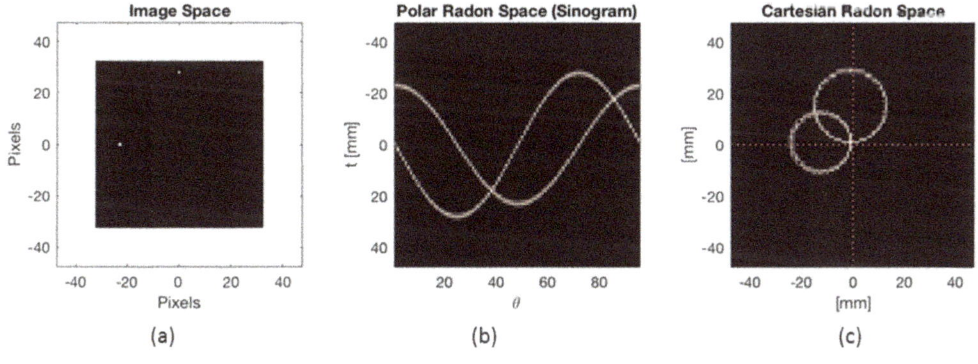

Figure 6.3. (a) Sample image and its Radon transform in (b) polar and (c) Cartesian coordinates.

Points on a line in the object space contribute to a single point in the Radon space. Each value of the function $R_\mu(x, y)$ accumulates contributions from all points in the object space $\mu(x, y)$.

In the three-dimensional object space, any function, e.g. $\mu(x, y, z)$, is described by a three-dimensional function $R_\mu(x, y, z)$ in the Radon space:

$$R_\mu(x, y, z) = \Re\{\mu(x, y, z)\}. \tag{6.5}$$

In 3D, the Radon hyperplanes are planes and the values of the 3D Radon transform $R_\mu(x, y, z)$ of a 3D function $\mu(x, y, z)$ represent superimposed *plane integrals* of μ. Similarly to the 2D case, each plane can be described by the coordinates of its distance vector. All planes that pass via point p in the object space are represented by locations in the Radon space lying on a spherical surface with a diameter defined by the origin of the coordinate systems and the coordinates of the point $\mu(x_p, y_p, z_p)$ (figure 6.4).

Each value of the function $R_\mu(x, y, z)$ accumulates contributions from all points from the object space $\mu(x, y, z)$.

The inverse Radon transform (IRT) recovers the values of the integrated function:

$$\mu(x, y, z) = \Re^{-1}\{R_\mu(x, y, z)\}. \tag{6.6}$$

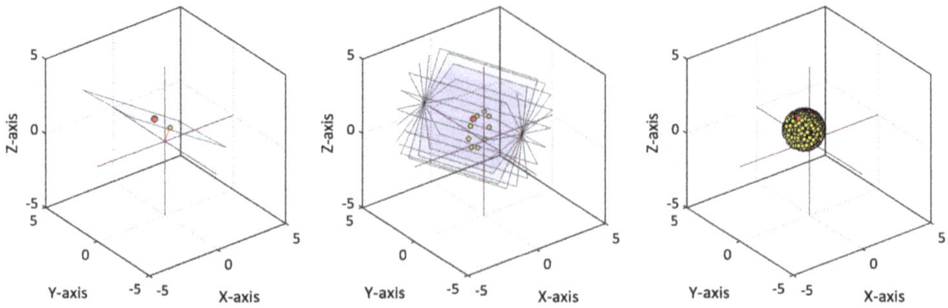

Figure 6.4. A point in the 3D object space (in red) and the closest to the origin of the coordinate system points (in yellow) of planes, passing through that point. The Radon points are shown for a single plane (a), for planes with coplanar normal vectors (b) and for even more, randomly oriented planes (c); planes in (c) are not shown for clarity.

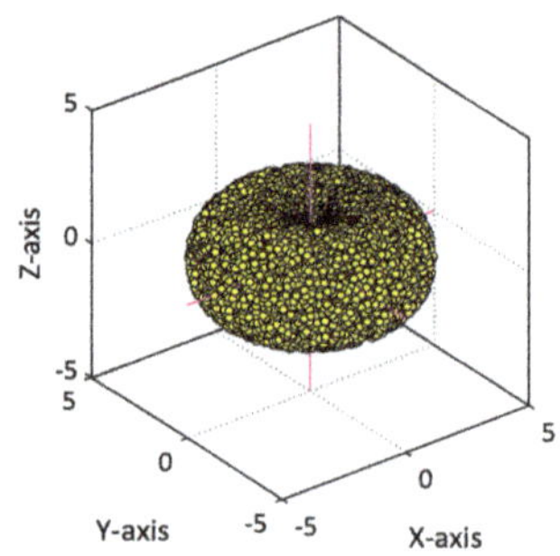

Figure 6.5. Randomly sampled points in the Radon space for a circular scanning trajectory.

To implement an inverse transform and reconstruct the object space, the values from the Radon space, to which the object space contributes, are needed.

6.2.6 Relation between Radon transform and CT

By taking a series of x-ray projections all around an object, one can compute line integrals of the linear attenuation coefficient $\mu(x, y, z)$, combine them into plane integrals, and eventually fill the Radon space.

In fact, what CT targets is the product of the IRT, i.e. to recover the values of $\mu(x, y, z)$ from its LIs, obtained by measurements according to equation (6.3). However, not every scanning set-up and trajectory can fill the Radon space. For example, one of the most popular scanning CBCT trajectories, the circular one, fills a toroidal area (figure 6.5) in the Radon space (Grangeat 1991).

Different computational approaches have been used and plenty of image reconstruction algorithms have been developed, described, and can be used for computing CT images. Two main types of algorithms can be distinguished—analytical and iterative. The analytical algorithms are based on strict mathematical descriptions, transforms, and function properties. The iterative algorithms can be algebraic or statistical. Algebraic algorithms consider the tomograms as matrices of unknown values and try to find them iteratively, exploiting error minimisation techniques. Statistical methods exploit models of the imaging systems and try to optimise the object content, so that the probability of the acquired projection data being reproduced is high. The first two types are briefly described in the following.

6.2.7. Image reconstruction

Analytical methods for image reconstruction

The methods in this group try to derive an analytical expression for computing the inverse Radon transform, i.e. computing the values of μ in the space. Based on the scanning geometry, and the properties of the x-ray source and the x-ray detector, different mathematical transformations are involved. The derivation is heavy but once done, the practical implementation still remains a challenge (e.g. additional factors influence the precision of the computations).

The Fourier slice theorem briefly states that any radial profile of the 2D Fourier transform of an image has the same values as the 1D Fourier transform of the LI projection of the same image obtained in a direction perpendicular to the direction of the profile (Kak and Slaney 1988). Figure 6.6 illustrates that.

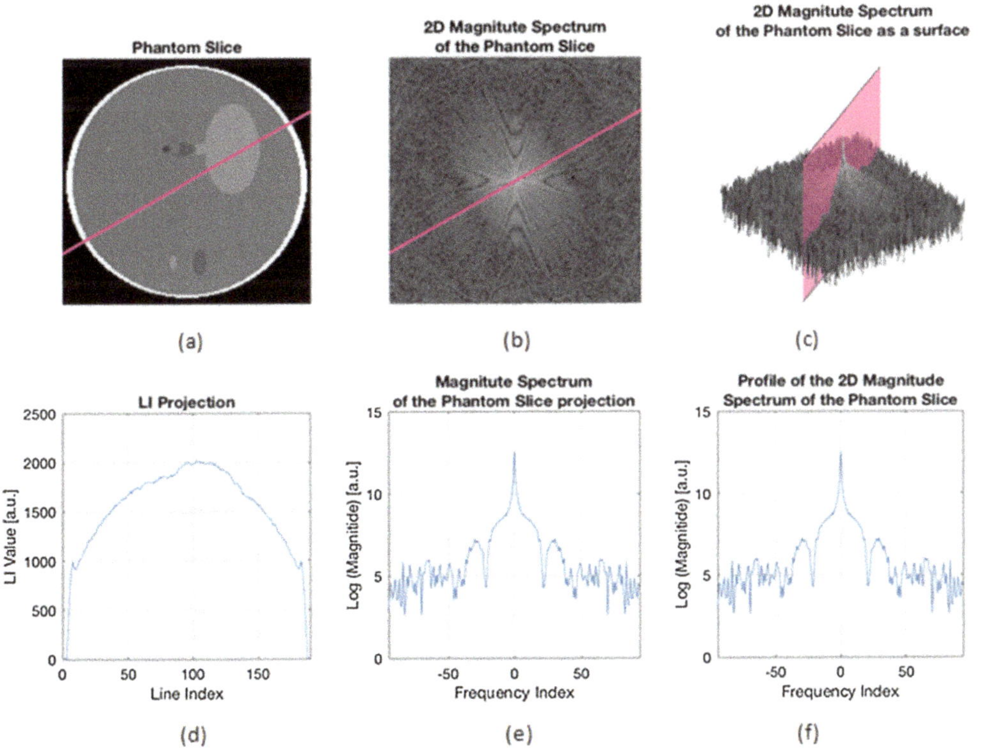

Figure 6.6. Illustration of the Fourier slice theorem: (a) the image, (d) its LI projection, (e) the 1D FT of the LI projection, (b, c) the 2D FT of the image, and (f) the corresponding profile of the 2D Fourier transform.

Direct Fourier methods rely on filling the 2D Fourier space from the spectra of the acquired projections using some type of interpolation (Gotlieb 2000). Working out the analytical expressions also leads to other practical implementations.

A classic example of the analytical methods is the algorithm of Feldkamp, Davis, and Kress (FDK) (Feldkamp *et al* 1984) for three-dimensional image reconstruction from cone-beam projections acquired along a circular scanning trajectory of the pair x-ray source–x-ray detector. Due to the planar motion of the source, the algorithm only provides exact reconstruction for the central axial plane, while for all other planes the reconstruction remains approximate. However, FDK has been popular for decades now due to the simplicity of its implementation—filtered back-projection (FBP).

The following example illustrates obtaining a set of projections, their high-pass filtered version, and the process of back-projection for reconstructing the unknown slice.

Example

A slice of the 3D Shepp–Logan phantom (Shepp and Logan 1974) (figure 6.7(a)) is artificially irradiated at each degree in the range [0°...180°], i.e. its sinogram (figure 6.7(b)) is obtained. Each column in the sinogram image is an LI projection of the

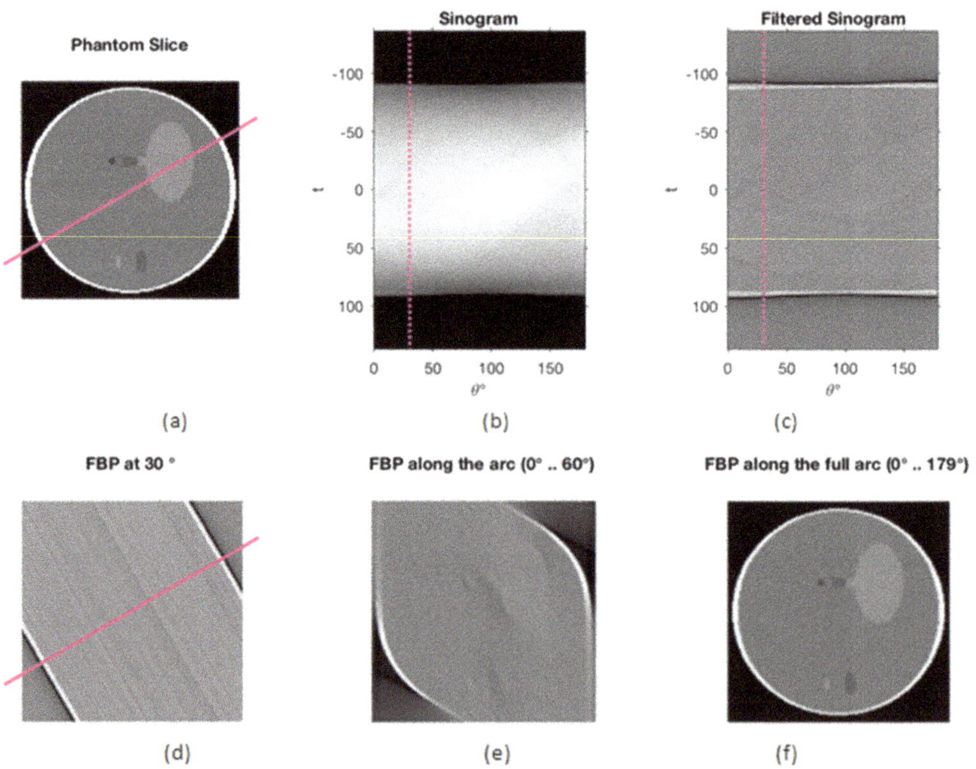

Figure 6.7. Illustration of filtered back-projection.

phantom slice from a different angle. Each projection is next filtered using a ramp-like high-pass filter according to the mathematics of the algorithm and the filtered sinogram is computed (figure 6.7(c)). The image reconstruction is done via back-projecting, i.e. smearing or distributing the values of each filtered projection along the direction in which they were acquired, and accumulating the result.

The bottom row of images in (figure 6.7) shows the result of a back-projection from one angle, an arc of angles, and the full arc of angles considering parallel x-ray beams.

An important work in the field of analytical 3D reconstructions is that of Grangeat (1991), which proposed an exact analytical FBP-type algorithm to compute the inverse three-dimensional Radon transform. The approach can also be used for other acquisition trajectories (Tam 1995).

Exact analytical reconstruction algorithms have also been derived for spiral scanning trajectories (Katsevich 2002).

One positive aspect of the analytical approach is the derivation and proving of a series of conditions and properties, which can then be taken into account in other types of reconstruction algorithms.

Algebraic methods for image reconstruction

Using computers, any volume of interest is represented as a three-dimensional matrix of *voxel* values $\mu(x_i, y_j, z_k)$. Voxels are the three-dimensional equivalents of pixels. Any projection task, e.g. obtaining an LI value, can be considered as building a specific weight sum of the voxel values along the corresponding beam line. If the voxel values are not known (as in the case of CT), then the CT scanning procedure can be considered as a process of building of a system of linear equations. The following example illustrates the approach.

Example

If a linear detector of four pixels is considered ($D_1 \ldots D_4$) and if a tomographic slice of also four unknown pixels ($V_1 \ldots V_4$) is targeted, then the scanning, or the obtaining of projection images of four different angles, can provide several equations:

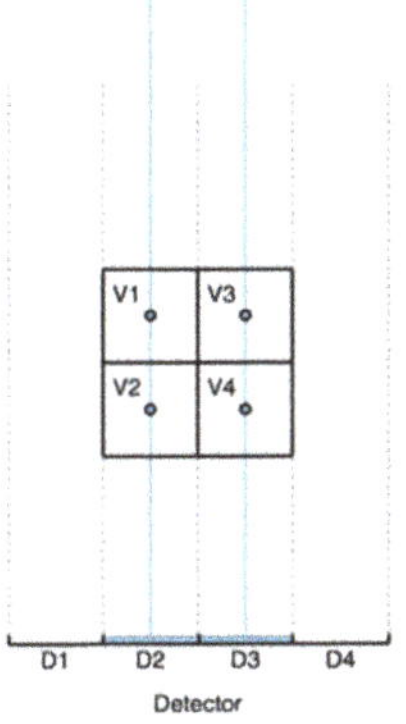


```
% Θ=0°, first index = 1
D(1,1) = 0;
D(1,2) = V(1)+V(2);
D(1,3) = V(3)+V(4);
D(1,4) = 0;
```

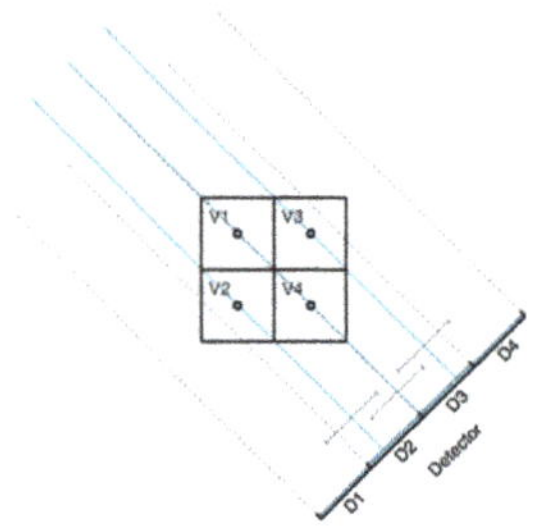


```
% Θ=45°, first index = 2
D(2,1) = 0.207*V(2);
D(2,2) = 0.500*V(1)+0.793*V(2)+0.500*V(4);
D(2,3) = 0.500*V(1)+0.793*V(3)+0.500*V(4);
D(2,4) = 0.207*V(3);
% 0.793 = 1.5-sqrt(2)/2
% 0.207 = sqrt(2)/2-0.5
```

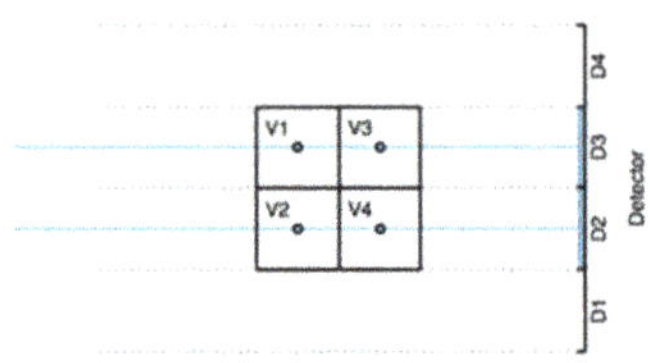


```
% Θ=90°, first index = 3
D(3,1) = 0;
D(3,2) = V(2)+V(4);
D(3,3) = V(1)+V(3);
D(3,4) = 0;
```

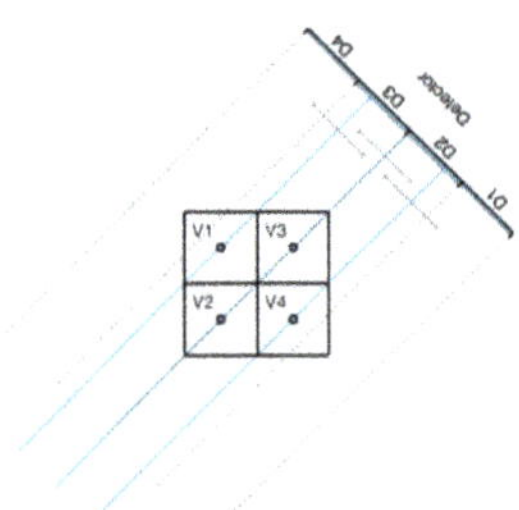


```
% Θ=135°, first index = 4
D(4,1) = 0.207*V(4);
D(4,2) = 0.500*V(2)+0.500*V(3)+0.793*V(4);
D(4,3) = 0.793*V(1)+0.500*V(2)+0.500*V(3);
D(4,4) = 0.207*V(1);
% 0.793 = 1.5-sqrt(2)/2
% 0.207 = sqrt(2)/2-0.5
```

If, for example, the unknown values $V_1 \ldots V_4$ were $V_1 = 1$, $V_2 = 2$, $V_3 = 3$, and $V_4 = 4$, then the system of equations would look like:

0.00000	0.00000	0.00000	0.00000		0.00000
1.00000	1.00000	0.00000	0.00000		4.00000
0.00000	0.00000	1.00000	1.00000		6.00000
0.00000	0.00000	0.00000	0.00000		0.00000
0.00000	0.20711	0.00000	0.00000		0.62132
0.50000	0.79289	0.00000	0.50000		4.87868
0.50000	0.00000	0.79289	0.50000		4.08579
0.00000	0.00000	0.20711	0.00000		0.41421
0.00000	0.00000	0.00000	0.00000		0.00000
0.00000	1.00000	0.00000	1.00000		7.00000
1.00000	0.00000	1.00000	0.00000		3.00000
0.00000	0.00000	0.00000	0.00000	V_1	0.00000
0.00000	0.00000	0.00000	0.20711	V_2	0.82843
0.00000	0.50000	0.50000	0.79289	$V_3 =$	5.67157
0.79289	0.50000	0.50000	0.00000	V_4	3.29289
0.20711	0.00000	0.00000	0.00000		0.20711

i.e.

$$\begin{bmatrix} A_{1,1}A_{1,2}A_{1,3}A_{1,4} \\ A_{1,1}A_{1,2}A_{1,3}A_{1,4} \\ \cdots \\ A_{12,1}A_{12,2}A_{12,3}A_{12,4} \end{bmatrix}_{12\times 4} \cdot \begin{bmatrix} V_1 \\ V_2 \\ V_3 \\ V_4 \end{bmatrix}_{4\times 1} = \begin{bmatrix} D_{1,1} \\ D_{1,2} \\ \cdots \\ D_{4,4} \end{bmatrix}_{12\times 1}.$$

The weights A_{ij} will depend on the location and the orientation of the beam, its diameter, and the distance between the beam and the voxels, and can be always prepared, i.e. they are known. The unknowns in the system of equations are the voxel values. The right-hand side of the equations is formed by the pixel values measured by the detector x-ray intensity.

Obviously, that reconstruction task will require solving of the system. The challenge here is the size of the system of equations. For a few equations a matrix division provides the result:

$$A \cdot V = D$$
$$A/(A \cdot V) = A/D$$
$$V = A/D.$$

However, for thousands of voxel values, this approach is not feasible due to the limited computational accuracy of the computer architectures and the accumulation of computational errors.

Iterative techniques provide a practical approach for solving the system of equations. The general idea can be summarised in several major steps:

1. Obtain a set of projection images.
2. Guess the initial slice content.
3. Project the guess and obtain a series of simulated projection images.
4. Compute the deviation between the real images and the simulated ones.
5. Check whether the obtained deviation is acceptable and if yes, exit!
6. Use the computed deviation to correct the initial guess for the slice image.
7. Go to 3.

Over the years, different smart approaches for accelerating the computations have been suggested and implemented, resulting in ART, SIRT, SART, MART and other algebraic reconstruction techniques (Gordon *et al* 1970, Kak and Slaney 1988).

Algebraic methods provide less noisy reconstructions but are computationally demanding. With the development of computer technology and parallel computing, their implementation is becoming feasible for larger tomographic images.

6.2.8 Errors and artefacts

The CT reconstruction algorithms are created assuming some sort of ideal conditions. In practice, this is never true. Various factors worsen the resolution of the reconstructions, and many other factors introduce artefacts and destroy the contrast of the tomograms. In the case of real CT reconstruction, the influence of such factors must be taken into account for the correct interpretation of the results. In the case of generating simulated projections for virtual CT studies, the influence of these factors must be additionally simulated to achieve a sufficient degree of realism. Some known factors that lead to noise and artefacts in the reconstructions are briefly presented in the following.

Random nature of x-ray generation and their interaction with matter. A series of random phenomena create noise in the projection images, such as: the random number of photons that leave the source; the random number of photons which do not interact, are scattered, or are absorbed while passing through the object; the random numbers of photon that are captured by the detector; the random number of visible light photons emitted by the scintillators for a captured x-ray photon, etc. The overall noise distribution can be approximated by a Poisson one (Macovski 1983). Therefore, lower x-ray exposure (mAs) settings lead, as a rule, to a higher level of inherent noise.

Characteristics of the components of the imaging chain. The size of the focal spot in the x-ray source is related to the penumbra effect of smoothing the images. Higher tube currents as in CT increase the spot size and amplify the effect (Bushberg 2002). The quality of the detector in terms of detector quantum efficiency and modulation transfer function affects the noise level and the frequency characteristics of the acquired images. Geometrical distortions are generally absent in flat panel detector images unless the detectors are displaced. High-sensitivity detectors as image intensifier tubes present typical non-linear geometrical distortions, however, and if used for reconstruction without corrections, this leads to significant blurring.

Geometrical set-up. The x-ray source and detector are mounted on a gantry or moved synchronously along specific scanning trajectories during the projection image acquisition. Imperfection in the mounting of the two components or incident displacements during the usage of the scanner, as well as non-precise trajectory implementation, lead to geometrical distortions and blurring.

Object properties. Large differences in the attenuation properties of the tissues of the structures of the image objects might lead to exceeding the dynamic range of detectors and saturation in the acquired values. Translated to the computed reconstructions this creates characteristic streak artefacts. As usually polychromatic x-ray sources are used in CT equipment, due to the fact that the linear attenuation coefficient of the tissues and materials is not constant for different x-ray energies and different attenuations, saturation might even only occur for specific energies. This phenomenon is known as beam hardening and also creates specific artefacts.

Reconstruction algorithms and applications. Iterative reconstructions can produce images with a better signal-to-noise ratio compared to images obtained using analytical methods for the same dose (Bushberg 2002). Better subjective and objective image quality is observed when applying iterative reconstructions in comparison to FBP-type reconstructions (Geyer *et al* 2015). The quality in FBP-types of algorithms is influenced by the reconstruction filter and the number of used projections. Using a small number of projections leads to characteristic stripes (aliasing) in the axial reconstructions (figure 6.8). The roll-off applied to the high-pass filter reduces the noise but worsens the spatial resolution of the reconstructions (figure 6.9).

The noise texture when using iterative methods is different. Applications based on approximate reconstructions using insufficient projection data can even have different artefact patterns.

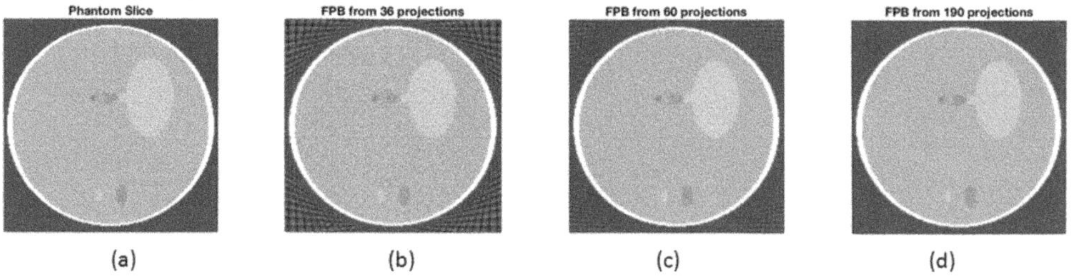

Figure 6.8. Parallel-beam FBP reconstruction of the Shepp–Logan phantom using a different number of projection images.

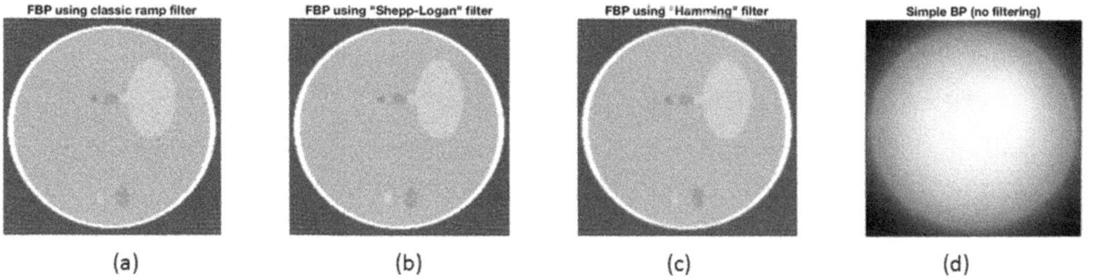

Figure 6.9. FBP reconstructions using different types of high-pass filters.

6.3 X-ray digital tomosynthesis

X-ray digital tomosynthesis (DTS), similarly to x-ray CT, produces tomographic images of the scanned objects. However, the reconstructions are usually computed from a small number of projection images, not providing enough data points in the Radon space. Thus, DTS usually requires short scanning trajectories and algorithms for approximate tomographic image reconstruction, which often are based on a kind of back-projection, e.g. shift-and-add (Dobbins and Godfrey 2003) or multiple projection algorithms (Kolitsi *et al* 1992), but iterative methods are also involved (Dobbins and Godfrey 2003).

A disadvantage of DTS comes from the fact that the projection data are limited and the reconstructions are approximate. This introduces a lot of artefacts and does not allow accurate quantitative measurements to be performed on the tomograms.

At the same time, DTS still provides the possibility to obtain 3D shape and location information on the structures within the scanned objects.

Different geometrical set-ups can be used to acquire projection images. Figure 6.10 depicts some of them.

Breast imaging benefits greatly from DTS. Digital breast tomosynthesis (DBT) is now routinely used in breast cancer screening (Vedantham *et al* 2015). Limited-arc partially isocentric scanning is usually applied and 11–50 images are acquired and processed. Although approximate, the 3D reconstructions provide valuable localisation information and help in the correct identification and classification of potential breast lesions and microcalcifications (El Bakry 2018). Non-destructive inspection can be mentioned as an example of a non-medical application of DTS.

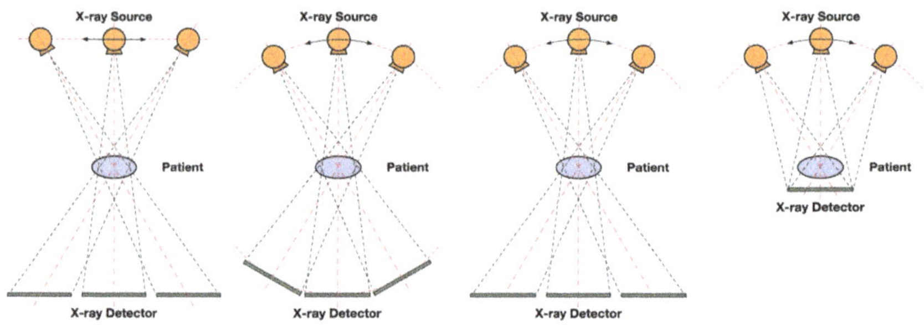

Figure 6.10. Examples of different DTS geometries: linear parallel-path (a), complete isocentric (b), partial isocentric (c), (d); (adapted from Dobbins and Godfrey 2003).

6.4 Image reconstruction in virtual studies

Physical and computational phantoms, including anthropomorphic ones, are valuable tools in various feasibility studies, providing preliminary and in-principle evaluation of new ideas, hypothesis testing, etc.

In the field of medicine, similar types of studies are usually referred to as virtual clinical trials (VCT). Appropriate anthropomorphic phantoms can serve to define or improve the quality assurance protocols for different modalities of medical imaging (DeWerd and Kissick 2014), including x-ray imaging, and are also of use for investigations on the limitations of current x-ray imaging techniques, as well as the design, testing, and evaluation of emerging (for instance phase-contrast CT, subtraction mammography, etc) x-ray imaging technologies and others.

6.4.1 Finding the right approach and tool

Performing research exploiting x-ray image reconstruction techniques obviously requires some preliminary knowledge—basic principles, requirements, challenges, etc. The first part of this chapter should provide a starting point to further reading and becoming familiar with the concepts.

The next step, performing reconstruction tasks, could be quite challenging for researchers who are not immediately familiar with the process. Finding the right approach and tool, reconstruction platform, or software development environment is a step forward. One can start by finding scripts for languages and environments such as *MATLAB* (https://mathworks.com/products/matlab.html) or *Octave* (https://octave-online.net/), others may prefer writing their own reconstruction software.

In the frame of the EUTEMPE course, the in-house developed *FDKR* tool is used illustrate the image reconstruction and the influence of different acquisition and scanning parameters on the tomogram image quality.

6.4.2 *FDKR*

FDKR has been created as a tool to help in the preliminary observation of projection images of different origin and their potential use for cone-beam CT or DTS image reconstruction. The image reconstruction is based on a back-projection. The tool also provides the possibility of converting the intensity image to line integrals and one-dimensional high-pass filtering, as required by the reconstruction algorithms.

Description

The main screen of the *FDKR* tool is shown in figure 6.11.

Four different panels can be seen: (1) *Monitor*, for displaying different images and set-ups; (2) *Projection Filtering*, for selecting the high-pass filter and optional filtering of the images; (3) *Cone-beam CT Reconstruction*, for setting up the reconstruction parameters; and (4) the *Status* window.

The Monitor

The *Monitor* is the place where different images or diagrams are shown. The content is selected with the help of the corresponding radio buttons in the *Monitor Selector* section. When displaying images, the monitor provides the possibility for grey-scale windowing by choosing the *User Signal Window* and using the two sliders below. The *Automatic Signal Window* scales the range of pixel values to the full range black–white.

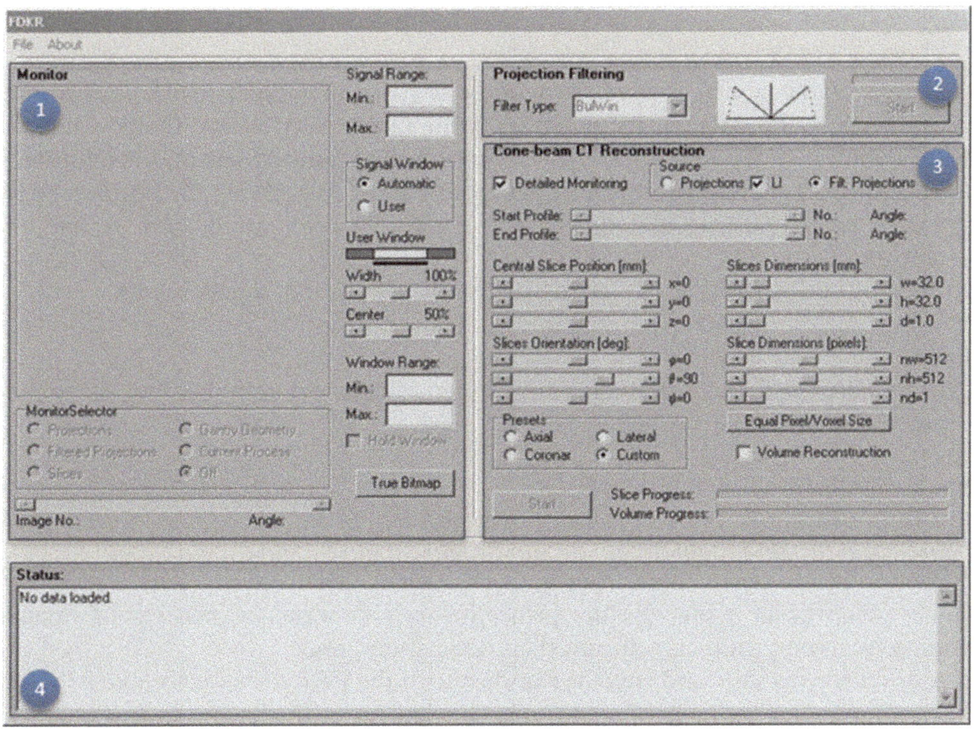

Figure 6.11. The main screen of *FDKR*.

Monitor Selector

The *Monitor* can be used to display five different types of images and scenes. The option *Projections* is used to browse the initial set of projection images. The filtered projection images, needed for classical tomographic reconstruction, can be produced in *FDKR* and then separately browsed using the next option of the *Monitor Selector* —*Filtered Projections*. The *Gantry Geometry* creates a proportional schematic representation of the geometrical set-up—source, detector, slice for reconstruction, and the main coordinate axes. When performing some processing, e.g. filtering or reconstruction, one can observe the temporary results if the option *Current Process* is selected. Finally, the result from the reconstruction can be observed by switching to *Slices*.

Projection Filtering

Filtered back-projection is a classical approach in computed tomography. According to the algorithm, each row in the initial projection images has to be filtered using a special high-pass filter—the ramp filter. Often, smoother versions of this filter are preferred in an attempt to reduce the noise level in the reconstructed images. Two types of filters can be selected and used for filtering the initial projection images. Clicking the *Start* button nearby invokes the filtering process. The filtering may be observed in real time if the *Monitor* is set to *Current Process* or, retrospectively, with the help of the slider under the *Monitor* if *Filtered Projections* has been selected.

Cone-beam CT Reconstruction

This area of the screen is used to define the types of images to be used for the reconstruction, the range of the images, i.e. the reconstruction arc, the position and the orientation of the tomogram to be reconstructed, and its size and resolution.

The group of four subgroups of sliders below is used to specify the position of the centre of the reconstructed tomographic image, its orientation, and its size and resolution.

The button for starting the reconstruction is also in this screen region.

Status

Various status and result messages are printed in this text area, helping the user to follow the activities performed.

Using the tool

Importing projection images

Projection images may come from various sources. The tool provides a simplified way to describe in a text file the projection sets in terms of number of images, naming, scanning geometry during their acquisition, etc.

A good way to start and become familiar with the *FDKR* tool is to take a look at the accompanying set of simulated projection images of the Shepp–Logan phantom (Shepp and Logan 1974) and its corresponding description file. The set should be found in a subfolder where the archive was unzipped. The description file usually has the extension .ASI and is created and modified with the help of any plain text editor, e.g. Notepad++ (https://notepad-plus-plus.org). The settings should be intuitively interpreted but a brief description is also provided after the following file example.

```
; ***** Gantry parameters and Reconstruction Settings *****
; ------------------------------------------------------------------------
SID               1000 ;[mm]
SIID              1300 ;[mm]
; ------------------------------------------------------------------------
; Projection Data
PJ_XSIZE          256      ;[mm]
PJ_YSIZE          256      ;[mm]
PJ_XLEN           256      ;[pixels]
PJ_YLEN           256      ;[pixels]
PJ_XOFFS          0        ;[mm]
PJ_YOFFS          0        ;[mm]
;
PJ_DRIVE          "C:"
PJ_DIR            "\FDKR\DATA\SL_HC_CIRC"
PJ_FNAMEBASEFMT   "real%d"
PJ_FNAMEEXT       ".flt"
PJ_HEADERSIZE     0  ;100
PJ_PRECISION      8  ;UINT8=1,UINT16=2,FLOAT=4,DOUBLE=8
PJ_STEP           1
PJ_START          0
PJ_END            360
PJ_ANGLEFILE      "C:\FDKR\DATA\SL_HC_CIRC\ANGLES.TXT"
;
WORK_DRIVE        "C:"
WORK_DIR          "\FDKR\DATA\SL_HC_CIRC\temp"
FIL_FNAMEEXT      ".fil"
;
; ------------------------------------------------------------------------
;
;            (O)   Source
;
;                   ^
;         ^ Z      / Y
;         |       /
;         |      /
;         |
;         | -----
;          /    /|    Volume
;         -----/ |
;         |    | |
;         |    |/
;         ------  --------> X
;
;           ^
;          / Y
;         /
;
;       -------------
;      /            /
;     /            /
;    /            /  Projection
;   --------------   ------------> X
;
```

The pseudo-graphical representation of the assumed geometry is usually provided at the end, which should also help to become oriented faster when preparing the content of the file before that.

Therefore, for any projection image set that is going to be used for reconstructions, a similar description file has to be prepared. The meaning of the individual keys is described below.

SID	Source–isocenter distance (mm)
SIID	Source–image_intensifier, i.e. source–detector distance (mm)
PJ_XSIZE	Width of the projection images (mm)
PJ_YSIZE	Height of the projection images (mm)
PJ_XLEN	Width of the projection images in pixels
PJ_YLEN	Height of the projection images in pixels

Normally the detector is expected to be 'centred', i.e. its geometrical centre is assumed to be in the path of the central x-ray beam and its rows to be parallel to the gantry rotation plane. In practice, it is difficult to achieve this and *FDKR* offers a possibility to account, at least, for the linear displacement of the detector:

PJ_XOFFS	Offset from the expected position along the detector's rows (mm)
PJ_YOFFS	Offset from the expected position along the detector's columns (mm)
PJ_HEADERSIZE	Number of bytes before the raw data in the image files
PJ_PRECISION	Pixel values repres. code, e.g. UINT8 = 1, UINT16 = 2, FLOAT = 4, DOUBLE = 8

The following keys specify the paths of the files from the projection set:

PJ_DRIVE	Drive letter, e.g. 'C:'
PJ_DIR	Path to the image files, e.g. '/FDKR/DATA/SL_HC_CIRC'

There are different naming conventions for the image files. Two of them are quite common: (i) including the gantry angle (in degrees) in the name of the file and (ii) using consecutive indices in the name of the files, while keeping track of the actual angles at which the projection images were taken, e.g. in a file. *FDKR* implements the second approach, but it may easily be tuned to work with the first one, as well. The following two keys define the format of the file names in terms of a base part and extension. C-like notation is used:

PJ_FNAMEBASEFMT	Image file naming in its base part, e.g. 'real%d'
PJ_FNAMEEXT	The extension in the image file naming, e.g. '.flt'

The following keys describe the values that will consecutively replace the '%d' part of the file name:

PJ_STEP	Step of the index for two consecutive images
PJ_START	First index
PJ_END	Last index

FDKR requires the presence of a text file, where each line contains the value of the angle at which the corresponding projection image in the sequence was taken. If angles are used for the naming, this file still has to be created, containing the same numbers as in the file names:

PJ_ANGLEFILE	Path and name of the angle file, e.g. 'C:\FDKR\DATA\SL_HC_CIRC\ANGLES.TXT'

The following keys define the working drive and the path for the files produced while using *FDKR*. This includes, of course, the filtered images, which by default use the same base name as the original ones but may have a different extension.

WORK_DRIVE	'C:'
WORK_DIR	'\FDKR\DATA\SL_HC_CIRC\temp'
FIL_FNAMEEXT	'.fil'

The definition file was intended to also contain some reconstruction parameters, but for the moment they are not processed. Even if you see anything related to this in the sample .ASI files, ignore it.

When the .ASI file is ready, select from the *FDKR* menu *File → Open Acquisition Data* and in the dialog that will appear select the file. If the images are correctly imported you will be able to browse them with the *Monitor* set in *Projections* mode. If not, something is still not provided correctly and the messages in the *Status* area can be used to determine and fix the problem.

Optional filtering
For the classical FDK algorithm, filtered projection images are needed. Therefore, for new sets, filtering has to be performed. In the *Projection Filtering* panel, choose the type of filter and click *Start*. All projection images will be filtered and stored under names and in a folder as specified in the .ASI file. Due to the linearity of the filtering operator, it does not depend on the concrete type of projection image (intensity or line integral images).

Selecting the reconstruction arc
The back-projection may be applied on any set of images. The FDK requires filtered line integral images along a 360° circular scanning path. DTS algorithms usually use a much shorter scanning arc. It can be selected with the help of the two relevant sliders.

Selecting the source images

By selecting the *Projections* option in the *Source* group of controls, the tool is instructed to use projection images for the reconstruction. The alternative is to use *Filt. Projections*. The tick box LI can be used to indicate that the initial images were line integral images and not intensity ones. This includes or excludes the additional internal transformation from intensity to line integral images before the reconstruction.

Selecting the tomogram location and orientation

What is expected to be defined is the coordinates (x,y,z) of the central point of a given slice of interest, and its orientation in terms of rotations around the global coordinate axes.

It is convenient while setting all these to have the *Monitor* in *Gantry Geometry* mode. The orientation is defined as a sequence of rotations around the z-, x-, and again z-axes at angles φ, θ, and ψ, respectively, starting parallel to the xOy plane orientation. Immediate selection of a central axial, sagittal, and coronary orientation of the reconstruction planes can be set using the *Presets* group of options.

Reconstruction

All the above details have to be correctly set before the reconstruction is run. Running of the reconstruction is done with the help of the nearby *Start* button. If the *Monitor* is in *Current Process* mode, one can observe the back-projection in real time.

Volumetric reconstruction may be an option. A stack of parallel slices is then reconstructed. The position and orientation the slice is defined around is considered to correspond to those of the centre of the stack slice. The corresponding tick box is used to request this, while the number of slices is chosen using the *nd* slider.

Grey-scale windowing for tomograms

By using the dedicated sliders, it is possible to select the visualised range of values.

True-size reconstructed images

After the reconstruction is completed, in the *Slice* mode of the *Monitor* it is possible to request a true-size version of the image (image pixels of the same size as the display's pixel) by clicking the relevant button. Then a bitmap of the image may be copied onto the clipboard for further use.

6.5 Self-training exercises

Exercise 1

The exercise aims to illustrate the difference between CT tomograms and DTS tomograms obtained with and without projection filtering. The Shepp–Logan (Shepp and Logan 1974) phantom is used as an object for imaging.

The following steps must be performed:

(a) Create the Shepp–Logan phantom as required in (section 4.3.2, exercise 4, chapter 4).
(b) Use the *XRAYImagingSimulator* to load the *Shepp_logan.phn* phantom.
(c) Visualise and obtain an idea of the phantom structures.
(d) Simulate the breast CT protocol set-up (simple irradiation) and produce projection images of the phantom:
   - SID = 1000 mm; SDD = 1300 mm.
   - Image resolution = 5 pixels/mm.
   - Image size: 800 × 800 pixels.
   - Acquisition arc: from 0° to 359°; step of 1°.
(e) Prepare an .ASI file for the generated projection image set.
(f) Import the projection set in FDKR.
(g) Produce a set of filtered images using the ramp-like filter.
(h) Perform three separate reconstructions of the central coronary plane of the phantom (after each of the reconstructions copy and save the result in a separate folder), using:
   - All filtered projections (CT).
   - Filtered projections along an arc of 40° (filtered DTS).
   - Non-filtered projections along the same arc of 40° (DTS).
(i) Compare the three tomograms visually (e.g. by opening them in *ImageJ*).

Exercise 2

This exercise aims to assess the potential of tomosynthesis to provide improved visibility of abnormalities in comparison to the classical radiography.

(a) Use the *XRAYImagingSimulator* and open *paraffin_water_spheres.phn* (prepared during exercise 2, chapter 4).
(b) Modify the phantom—change the property of one of the spheres to $CaCO_3$. Change its radius to 0.1 mm.
(c) Visualise the phantom and remember the position of the $CaCO_3$ sphere.
(d) Generate a planar image of the phantom, using the following settings, and save the image in a folder *Exercise6_2a*:
   - Energy of the x-rays: 19 keV.
   - Acquisition angle: 0:2:0 deg.
   - SID: 500 mm; SDD: 550 mm.
   - Image resolution: 10 pixels/mm; image size: 700 × 700.
(e) Use the *XRAYImagingSimulator* to simulate 21 projection images in tomosynthesis mode. For this purpose, use the *Analytical Image Simulation* menu and the following settings:
   - SID = 500 mm; SDD = 550 mm.
   - Acquisition angle: −200:20:200.
   - Image size = 700 × 700 pixels; image resolution = 10 pixel/mm.
   - Folder: *Exercise6_2b*.

(f) Reconstruct images using *FDKR*.
(g) Visually compare tomograms and planar images.
Write a summary of the study.

References

Bushberg J T 2002 *The Essential Physics of Medical Imaging* (Philadelphia, PA: Lippincott Williams & Wilkins)

DeWerd L and Kissick M 2014 *The Phantoms of Medical and Health Physics* (New York: Springer)

Dobbins J T and Godfrey D J 2003 Digital x-ray tomosynthesis: current state of the art and clinical potential *Phys. Med. Biol.* **48** R65–R106

El Bakry R A R 2018 Breast tomosynthesis: a diagnostic addition to screening digital mammography *Egypt. J. Radiol. Nucl. Med.* **49** 529–35

Feldkamp L A, Davis L C and Kress J W 1984 Practical cone-beam algorithm *J. Opt. Soc. Am. A* **6** 612–9

Geyer L L *et al* 2015 State of the art: iterative CT reconstruction techniques *Radiology* **276** 338–56

Gordon R *et al* 1970 Algebraic reconstruction techniques (ART) for three-dimensional electron microscopy and x-ray photography *J. Theor. Biol.* **29** 471–81

Gottlieb D, Gustafsson B and Forssén P 2000 On the direct Fourier method for computer tomography IEEE Trans *Med. Imag.* **19** 223–32

Grangeat P 1991 Mathematical framework of cone beam 3D reconstruction via the 1st derivative of the radon-transform *Mathematical Methods in Tomography* Lecture Notes in Mathematics vol 1497 (New York: Springer) pp 66–97

Kak A C and Slaney M 1988 *Principles of Computerized Tomographic Imaging* (Piscataway, NJ: IEEE)

Katsevich A 2002 Theoretically exact filtered backprojection-type inversion algorithm for spiral CT *SIAM J. Appl. Math.* **62** 2012–26

Kolitsi Z *et al* 1992 A multiple projection method for digital tomosynthesis *Med. Phys.* **19** 1045–50

Macovski A 1983 *Medical Imaging Systems* (Englewood Cliffs, NJ: Prentice-Hall)

Natterer F 1986 *The Mathematics of Computerized Tomography* (Wiesbaden: Vieweg and Teubner)

Shepp L A and Logan B F 1974 The Fourier reconstruction of a head section *IEEE Trans. Nucl. Sci.* **21** 21–43

Tam K C 1995 Exact image reconstruction in cone beam 3D CT *Rev. Prog. Quantit. Nondestruct. Eval.* **14** 657–64

Vedantham S *et al* 2015 Digital breast tomosynthesis: state of the art *Radiology* **277** 663–84

IOP Publishing

Anthropomorphic Phantoms in Image Quality and Patient Dose Optimization
A EUTEMPE Network book
Kristina Bliznakova, Ivan Buliev and Zhivko Bliznakov

Chapter 7

Virtual studies with anthropomorphic phantoms

7.1 Introduction

The potential area of application of anthropomorphic phantoms is as vast as the whole variety of clinical routine examinations, research, and clinical engineering procedures. The aim of this chapter is, by giving selected examples of virtual case studies, to transfer the basic idea of using the phantoms for solving different problems. These include the design and evaluation of advanced imaging techniques as well as the design of anthropomorphic phantoms for imaging and dosimetry purposes. Two virtual case studies have been selected as examples:

- The use of software phantoms in the design and evaluation of advanced x-ray imaging techniques: the case of dual-energy imaging;
- Use of software phantoms in the design and evaluation of advanced x-ray imaging techniques: the case of phase-contrast imaging.

Self-training video tutorials are available at the following links:

http://tuv.eutempe-net.eu/wp-content/uploads/Education-materials/Videos/Videomamo-1%20optimized.mp4.

http://tuv.eutempe-net.eu/wp-content/uploads/Education-materials/Videos/Phantom%20Lucmfr-1%20optimized.mp4.

http://tuv.eutempe-net.eu/wp-content/uploads/Education-materials/Videos/Phantom%20Scan-1%20optimized.mp4.

These examples are similar to the projects that participants will develop during the face-to-face phase of the EUTEMPE course in Varna. The preliminary familiarisation with the application of the phantoms, the tutorials, and the practical work during the face-to-face session, are expected to provide the required knowledge and skills for successful project development.

doi:10.1088/2053-2563/aae197ch7

7.2 Anthropomorphic phantoms for design and evaluation of advanced x-ray imaging techniques: practical examples

7.2.1 Applications of anthropomorphic phantoms for design and evaluation of dual-energy imaging

7.2.1.1 Theory

The concept of using two photon energies to obtain information on tissue characteristics was first suggested by Jacobson (1953). In dual-energy imaging, two images are acquired at different energies. The nonlinear dependence of the absorption coefficients with x-ray energy allows a weighted subtraction of the original images, resulting in removing one type of tissue from the image, while others are improved.

Methods used to reconstruct the dual-energy image vary from nonlinear to straightforward logarithmic dual-energy subtraction techniques (Johns *et al* 1983, Johns *et al* 1985, Johns and Yaffe 1985, Asaga *et al* 1987, Boone and Shaber 1990, Marziani *et al* 2002, Taibi *et al* 2003, Badr *et al* 2014, Taibi and Vecchio 2014). Here, the second approach is presented, where two images are acquired at two different *low* and *high* energies and then subtracted properly in order to obtain information for tissue characteristics. In the case of breast, the following assumption is made. The breast may be considered as being composed of three different components: glandular, adipose, and cancerous tissue. To simplify the modelling task further, the adipose and glandular tissue may be represented as one tissue with proper adipose-glandular weighted contributions and to have two-component breast structures: namely breast tissue and calcification. This is reflected in figure 7.1, where the compressed breast is composed of breast tissue and an abnormality with *c* (from calcification) thickness.

Using the dual-energy technique, the contrast due to the distribution of normal breast tissues is expected to be suppressed while at the same time the intrinsic contrast of the breast lesion is enhanced. The result of the algorithm is in principle a calcium image (denoted as *CI*), assuming that the breast is composed of breast tissue and calcium abnormalities. The use of this method focuses on the detection of small calcifications rather than on the tissue characterisation.

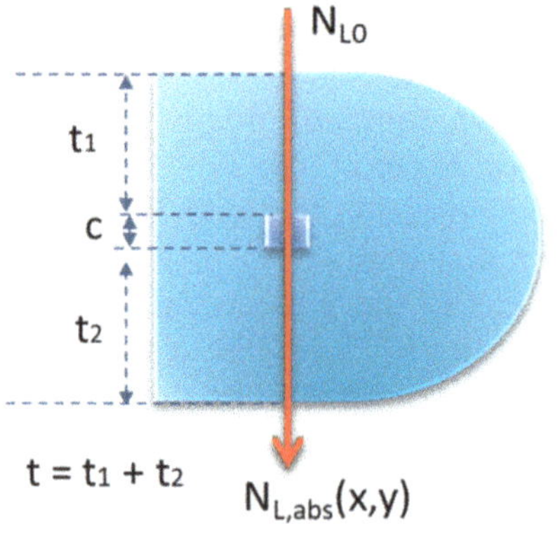

Figure 7.1. A sketch demonstrating the basics of dual-energy imaging.

Let $t(x,y)$ and $c(x,y)$ represent the breast tissue and calcification thicknesses over the image plane (x,y) and let μ_L and μ_H represent the effective linear attenuation coefficients for the breast tissue for low- and high-energy images. The low- and high-energy absorbed fluences can be approximated by:

$$\begin{aligned} N_{L,abs}(x, y) &= A_L N_{L0}\exp(-\mu_L t(x, y) - \xi_L c(x, y)) \\ N_{H,abs}(x, y) &= A_H N_{H0}\exp(-\mu_H t(x, y) - \xi_H c(x, y)), \end{aligned} \tag{7.1}$$

where $N_{H,abs}$ and $N_{L,abs}$ are the number of absorbed photons for the respective detector for low and high energies. A_L and A_H are the absorption fractions of the detector for these low and high energies, while N_{LO} and N_{HO} are the incident un-attenuated photon beam. Since screens are considered energy detectors, the images formed from the above photon fluences are:

$$\begin{aligned} I_H(x, y) &= N_{H,abs}(x, y)E_{H,abs}(x, y) \\ I_L(x, y) &= N_{L,abs}(x, y)E_{L,abs}(x, y), \end{aligned} \tag{7.2}$$

where $E_{H,abs}$ and $E_{L,abs}$ are the average energy absorbed by a photon for these low and high energies, respectively. All operations are applied on a pixel by pixel basis (denoted as (x,y)). The logarithm of equation (7.2) results in:

$$\begin{aligned} \ln[I_H(x, y)] &= \ln[N_{H,abs}(x, y)E_{H,abs}(x, y)] \\ &= \ln[E_{H,abs}(x, y)A_H N_{H0}(x, y)\exp(-\mu_H t(x, y) - \xi_H c(x, y))] \\ \ln[I_L(x, y)] &= \ln[N_{L,abs}(x, y)E_{L,abs}(x, y)] \\ &= \ln[E_{L,abs}(x, y)A_L N_{L0}(x, y)\exp(-\mu_L t(x, y) - \xi_L c(x, y))], \end{aligned} \tag{7.3}$$

which is easily transformed to the following equation:

$$\begin{aligned} \ln[I_H(x, y)] &= \ln[E_{H,abs}(x, y)A_H N_{H0}(x, y)] \\ &\quad + (-\mu_H t(x, y) - \xi_H c(x, y)) \\ \ln[I_L(x, y)] &= \ln[E_{L,abs}(x, y)A_L N_{L0}(x, y)] \\ &\quad + (-\mu_L t(x, y) - \xi_L c(x, y)). \end{aligned} \tag{7.4}$$

Equation (7.4) is equivalent to:

$$\begin{aligned} \mu_H t(x, y) + \xi_H c(x, y) &= \ln[E_{H,abs}(x, y)A_H N_{H0}] - \ln(I_{H,abs}(x, y)) \\ \mu_L t(x, y) + \xi_L c(x, y) &= \ln[E_{L,abs}(x, y)A_L N_{L0}] - \ln(I_{L,abs}(x, y)). \end{aligned} \tag{7.5}$$

By multiplying the second equation of (7.5) with $R = \frac{\mu_H}{\mu_L}$, the following result is obtained:

$$\begin{aligned} \mu_H t + \xi_H c &= \ln[E_{H,abs} A_H N_{H0}] - \ln(I_{H,abs}) \\ \mu_H t + R\xi_L c &= R\ln[E_{L,abs} A_L N_{L0}] - R\ln(I_{L,abs}), \end{aligned} \tag{7.6}$$

where for simplicity the notation for the pixel (x,y) is omitted. Equation (7.6) is rearranged as follows:

$$\begin{aligned} \mu_H t + \xi_H c &= \ln[E_{H,abs} A_H N_{H0}] - \ln(I_{H,abs}) \\ -\mu_H t - R\xi_L c &= -R\ln[E_{L,abs} A_L N_{L0}] + R\ln(I_{L,abs}). \end{aligned} \tag{7.7}$$

The result of the sum between the two equations in (7.7) is:

$$\xi_H c - R\xi_L c = \ln[E_{H,abs} A_H N_{H0}] - \ln(I_{H,abs}) - R\ln[E_{L,abs} A_L N_{L0}] + R\ln(I_{L,abs}). \tag{7.8}$$

The left part of equation (7.8) is called the calcium image (CI) and is represented as:

$$CI = c(R\xi_L - \xi_H) = \ln(I_{H,abs}) - R\ln(I_{L,abs}) + k, \tag{7.9}$$

with k given as:

$$k = \ln\frac{[E_{L,abs} A_L N_{L0}]^R}{[E_{H,abs} A_H N_{H0}]}. \tag{7.10}$$

7.2.1.2 Tutorial

The purpose of this tutorial is to demonstrate how to implement a dual-energy virtual imaging study using a simple computational phantom. The following tasks are carried out in order to generate the dual-energy image:

(a) Load the CSG phantom *paraffin_phantom.phn* generated in chapter 4, section 4.3.2, exercise 1, i.e. the paraffin phantom with three water spheres.
(b) Simulate a radiography image at 20 keV; SID = 1000 mm; SDD = 1300 mm; pixel size = 10 pixels/mm; image size = 1000 pixels × 1000 pixels. Use the option *Image formation*. The obtained simulated image is a line integral: μx.
(c) Simulate one radiography image at 30 keV; SID = 1000 mm; SDD = 1300 mm; pixel size = 10 pixels/mm; image size = 1000 pixels × 1000 pixels. Use the option *Image formation*. The simulated image is a line integral: μx.
(d) Multiply the low-energy images by −1 and obtain the following: $-\mu x$ ($\ln(I) = -\mu x$).
(e) Suppose ideal images, no noise added.
(f) Calculate the R value.
(g) Multiply the low-energy image by R.
(h) Sum the low- and high-energy images by using *ImageJ* or *MATLAB/Octave,* for instance.

$$R = \frac{\mu_{\text{background}(E_H)}}{\mu_{\text{background}(E_L)}} = \frac{0.2520}{0.4010} = 0.6285.$$

Figure 7.2 depicts the implementation of the above (a)–(h) steps. Figure 7.2(a) shows the original low-energy image obtained after executing steps (a) and (b). Then, the pixel values of this image represent a line integral. These values are multiplied by −1 (step (d)) and by R (steps (f) and (g)) and the resulting image is shown in figure 7.2(b). Figure 7.2(c) shows the high-energy image obtained as a result of the execution of step (c). The calculated *dual-energy* image is shown in figure 7.2(d) as a result of step (h).

7.2.1.3 Virtual study

The aim of this study is to use a computational breast phantom to perform a virtual study to determine the potential of dual-energy mammography for improved visibility of breast abnormalities, compared to single-energy mammography.

Methods:

- *Phantom(s)*: LUCMFR—small to medium size, average compressed breast thickness, texture realised as spheres ranging between 1.0 mm and 15.0 mm in diameter.
- *Abnormalities*: One mass abnormality and one individual microcalcification.
- *Imaging techniques*: Dual-energy mammography, monoenergetic beams.
- *Any other parameters* are at your discretion.

Tasks:

(a) Create a phantom according to the provided specification and add the abnormalities. Choose appropriate materials from table 4.1, chapter 4.
(b) Obtain synthetic conventional mammographic images of the phantom at low energy in the range 17–23 keV, and at high energy in the range 50–80 keV, selected at your discretion. Add photon noise to images.
(c) Compute dual-energy difference images.
(d) Compare the images subjectively (visually).
(e) Compute the objective figures of merit (FOMs)—contrast in the region of abnormalities and compare.

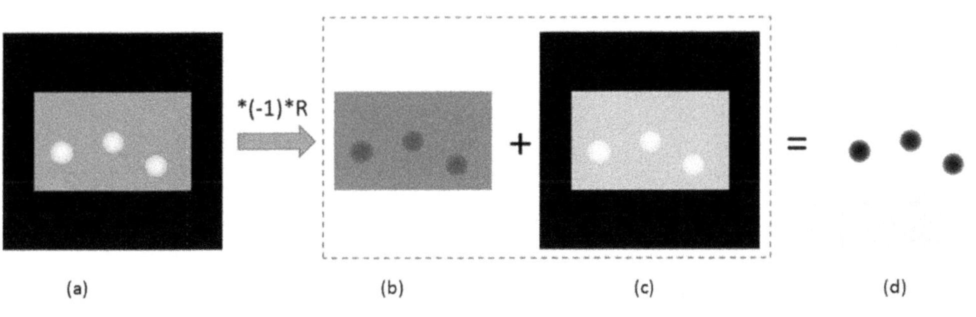

Figure 7.2. Dual-energy imaging demonstration: (a) low-energy image, (b) $(-R)^*$ low-energy image, (c) high-energy image, and (d) calcification image (CI).

To be prepared:

- *A report*: a short written summary of the work and the obtained results; containing for example:
    - appropriate outcome images and profiles,
    - tables with computed FOMs,
    - discussion of the results and conclusions.
- *A PowerPoint presentation*: to be used for a short (10 minute) oral presentation.

Useful hints:

- The tool for phantom creation is the *LUCMFRGen* tool.
- For abnormalities, change the properties of two of the spheres by using the *XRAYImagingSimulator* tool.
- Projection images in absorption mode are generated using the *XRAYImagingSimulator* tool. Have the *LUCMFR case* option checked on as well as *Fast image formation.*
- Calculate dual-energy projection images by using *ImageJ* or *Octave/ MATLAB.*

***Implementation* of the project**

Step 1*:* Building the phantom. The *LUCMFRGen* tool (figure 7.3(a)) was used to generate the phantom, shown in figure 7.3(b). The tissue-equivalent materials chosen for the phantom are summarised in table 7.1.

Step 2*:* Image formation simulation is implemented as shown in figure 7.3(b) and table 7.2. One low-energy and one high-energy image are acquired. Photon noise is added to images. The incident exposures to the phantom were set to 2 mGy and 1 mGy, for low- and high-energy beams, respectively. Images are shown in figure 7.4.

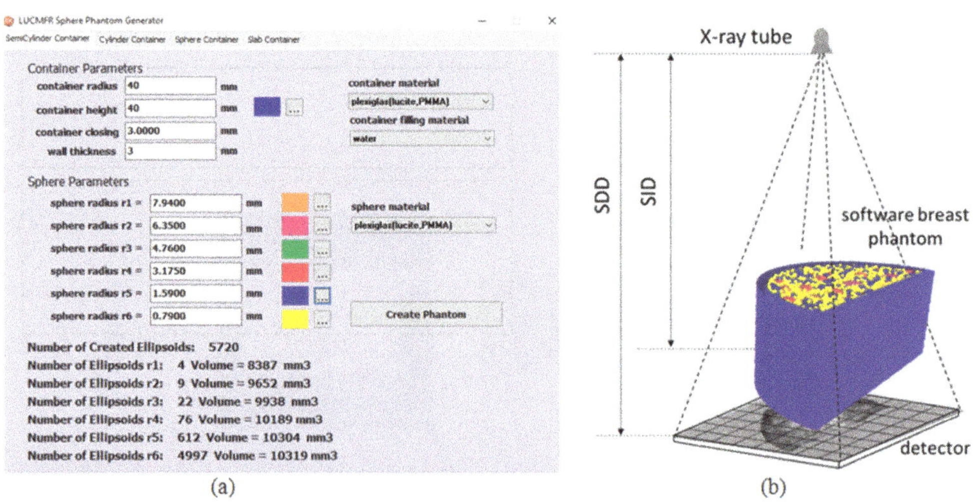

Figure 7.3. Generation of the computational phantom: (a) parameters used for the creation of the phantom and (b) the phantom and the image formation simulation set-up.

Table 7.1. Parameters of the computational phantom.

Name	Shape	Dimensions	Material
Container	Semi-cylinder	r = 40 mm, h = 40 mm, wall thickness = 3 mm	PMMA
Texture	Spheres	r = 7.95, 6.35, 4.76, 3.18, 1.59, 0.79 mm	PMMA
Filling	—	—	Air
Mass	Sphere	r = 1.5 mm	100% gland
μCs	Sphere	r = 0.07 mm	Hydroxyapatite

Table 7.2. Parameters of the imaging system used in the simulation.

Parameter	Value
SID	600 mm
SDD	650 mm
Detector resolution	100 μm
Image size	1000 ×1000 pixels
Low energy	19 keV
High energy	50 keV

Table 7.3. Linear attenuation coefficients, cm^{-1}.

keV	μ_{PMMA}, cm^{-1} (ρ = 1.19 g cm^{-3})	$\mu_{hydroxy}$, cm^{-1} (ρ = 2.22 g cm^{-3})
19	0.756	7.595
50	0.247	0.5867

Step 3: Calculation of dual-energy images. Image formation simulation is implemented as shown in figure 7.3(b) and table 7.3. Our study will be limited to two low energies and two high energies. Simulated projection images are shown in figures 7.4(a), (b).

R was calculated according to equation (7.11): $R = \frac{\mu_H}{\mu_L} = 0.327$. The calculated dual-energy images with no photon noise initially added to the images and with noise are shown in figures 7.4(c), (d), while their magnified versions are depicted in figures 7.4(e), (f). Visually, one can observe the improved visibility of both the small μCs and the mass in the dual-energy images in comparison to both the low- and high-energy images. Objective evaluation is not accomplished, due to the fact that both abnormalities are not visible in the low-energy image.

This is one example of a work project assignment. During the face-to-face session, the participants will recreate and use this phantom with a different composition and abnormalities and will perform various virtual clinical studies in mammography, tomosynthesis, CBCT, and phase-contrast imaging.

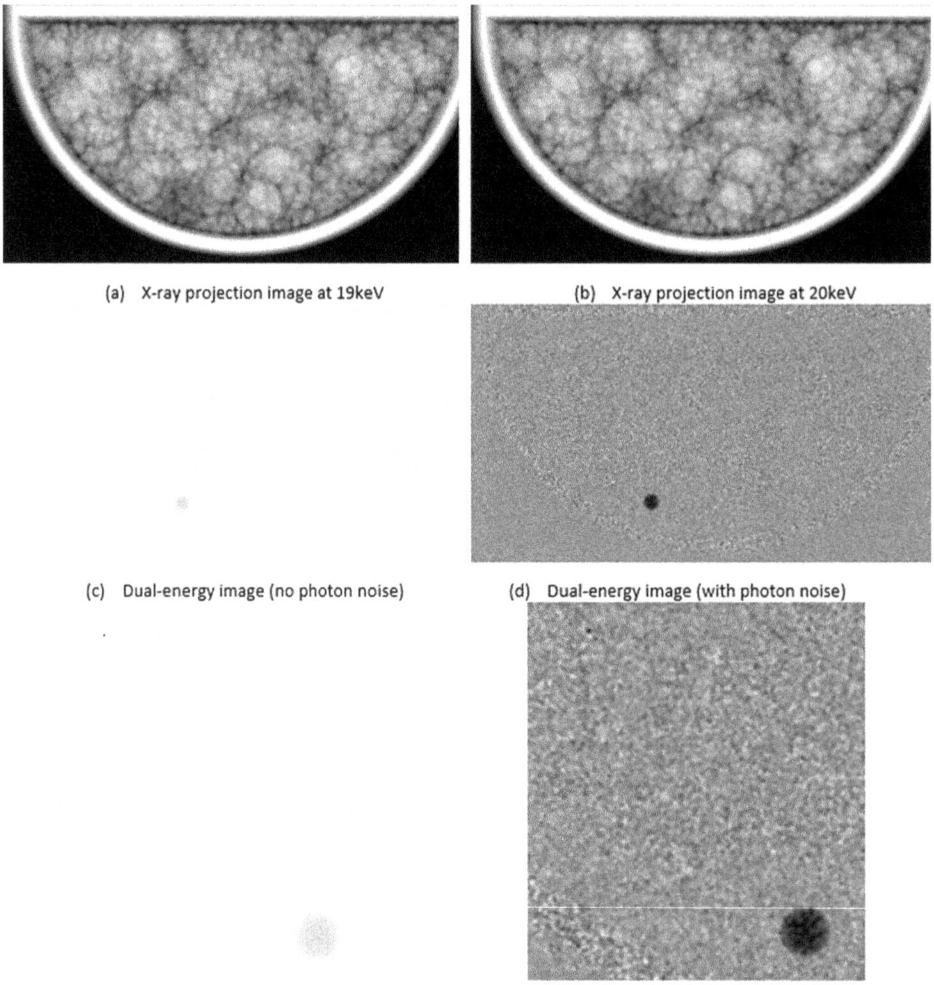

(a) X-ray projection image at 19keV

(b) X-ray projection image at 20keV

(c) Dual-energy image (no photon noise)

(d) Dual-energy image (with photon noise)

(e) ROI from the dual-energy image (no noise)

(f) ROI from the dual-energy image (with noise)

Figure 7.4. Simulated images: (a, b) x-ray projection images, (c, d) calculated dual-energy images, and (e, f) regions of interest (ROI) from the dual-energy images.

7.2.2 Applications of anthropomorphic phantoms for design and evaluation of phase-contrast imaging

7.2.2.1 Theory

During the last decade, a number of novel methods have been developed in the field of x-ray imaging, based on the wave properties of x-rays. When a wave interacts with matter both its amplitude and phase are modified. This change can be described by a complex form of the *refractive index* n of the medium, the value of n deviating only very slightly from unity. Inside the medium with refractive index $n = 1 - \delta + i\beta$ the wave becomes phase shifted with respect to the wave propagating in free space, as shown in figure 7.5. Here, δ is the refractive index decrement due to phase shift,

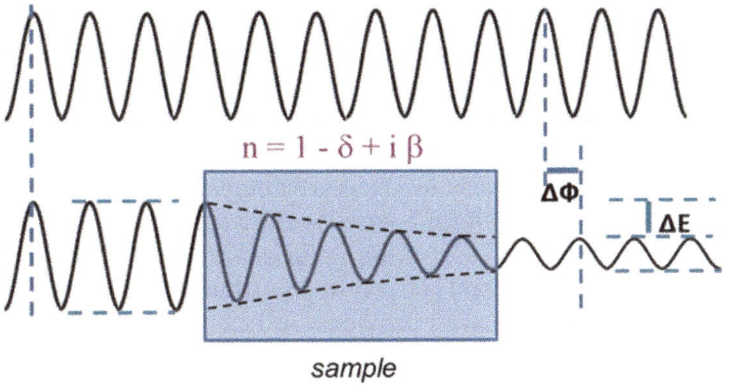

Figure 7.5. Wave propagation through a sample.

while β describes x-ray absorption. This allows an increase in image contrast, especially in imaging weakly absorbing samples (Arfelli *et al* 1998, 2000, Lewis 2004).

Several methods have been developed to obtain x-ray phase-contrast images, which can be classified into x-ray interferometry, diffraction-enhanced imaging, in-line phase-contrast imaging, grating interferometry, and coded-aperture techniques. Some of these x-ray phase-contrast imaging methods require highly monochromatic plane wave radiation and sophisticated x-ray optics. Examples are the first two methods above, which require a rather complex experimental set-up compared to the third method, known as propagation-based, which is a relatively simple method to implement, essentially requiring only a small focus x-ray tube and electronic detection.

XRAYImagingSimulator provides the possibility to simulate the propagation-based method (Bliznakova *et al* 2015). The method for obtaining phase-contrast images is shown in figure 7.6. The simulation of phase-contrast imaging in this

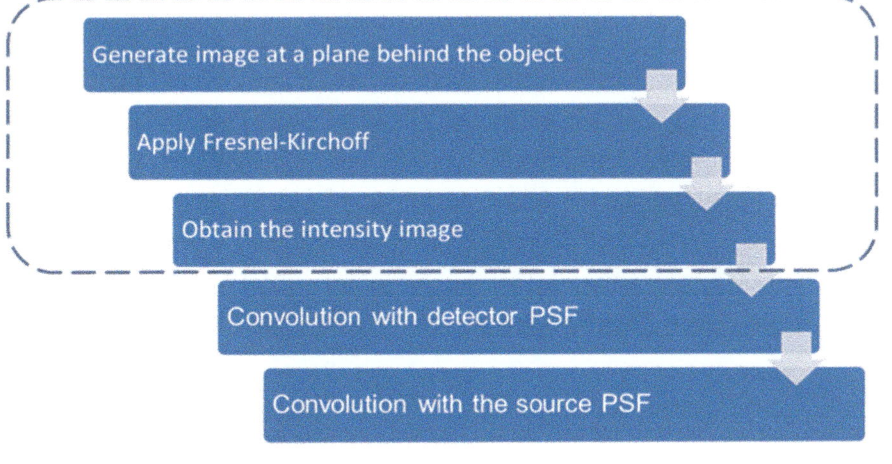

Figure 7.6. Generation of phase-contrast images.

software tool is based on the Fresnel–Kirchhoff diffraction theory. Specifically, it considers pure coherent radiation emerging from a point monochromatic source, located at a distance Z_{so} from the object plane. The object is placed in an object plane that is at $z = 0$ and the detector is located at a distance Z_{od} from the object plane.

The free space propagation technique has advantages concerning the simplicity of the experimental set-up. It requires no optics to form images as at x-ray wavelengths lenses are difficult to manufacture. In-line phase-contrast x-ray imaging provides images where both absorption and refraction contribute. In order to achieve phase-contrast imaging the x-ray tube must provide a sufficient degree of lateral coherence, given as:

$$\frac{\lambda Z_{so}}{s}, \tag{7.11}$$

where λ is the wavelength, Z_{so} is the distance between the source and the object, and s is the focal spot size of the tube. Therefore, phase information retrieval is easier with low-energy x-ray photons, a large source-to-object distance, and a small source size. Temporal coherence or monochromaticity is not essential and, as such, phase-contrast imaging can also be performed with polychromatic radiation.

Figure 7.7 shows a comparison between two acquisition geometry arrangements: the geometry in (a) is similar to the traditional absorption x-ray imaging, where the detector is placed close to the patient, while the geometry in (b) is dedicated to phase-contrast, where the detector is placed at a large distance from the patient. In figure 7.7, the x-ray beam passes through the edges of a cylinder of tissue-equivalent

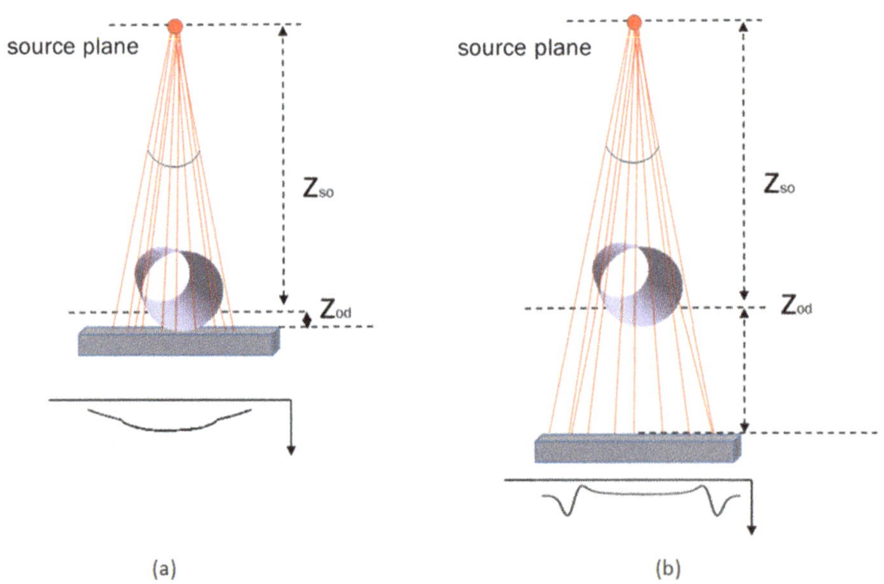

Figure 7.7. Imaging in (a) the absorption and (b) free propagation phase-contrast modes.

material. The x-rays refract from the edge and the corresponding beam deviation by the edge region produces a sharp fringe pair on the detector that results in improvement in the edge visibility. This edge will be very well visualised at a sufficient distance between the object and the detector. This distance gives time for the phase introduced by the sample to develop into phase-contrast.

7.2.2.2 Tutorial

The purpose of this tutorial is to demonstrate how to implement a phase-contrast virtual imaging study using a simple nylon wire. In particular, the aim is to investigate the influence of three different object-to-detector distances on the image quality of the projection image: SDD = 1001 mm, 1300 mm, and 2300 mm (figure 7.8).

Tasks

The first subtask is to create the nylon wire as a cylinder with a diameter of 200 μm. For this purpose, the *XRAYImagingSimulator* is used. The Euler angels (φ, θ, ψ) are set to (90°, 90°, 0°). The cylinder is placed at the isocentre point (0, 0, 0).

Once the phantom is created, the *Analytical Image Simulation* → *Image Simulation—solid geometry* is open.

The set-up shown in figure 7.8(a) is for the absorption mode. In this case, the detector is placed close to the object. To implement this set-up, the detector is placed at a distance of 1001 mm away from the source, while the source to isocentre distance is 1000 mm, i.e. the object to detector distance is equal to 0.9 mm. The image resolution is set to 200 pixels per mm, which corresponds to a pixel size of 0.5 μm. Projection images and profiles taken across them for the three distances 1001 mm, 1300 mm, 2300 mm are shown in figure 7.9. The improvement of object

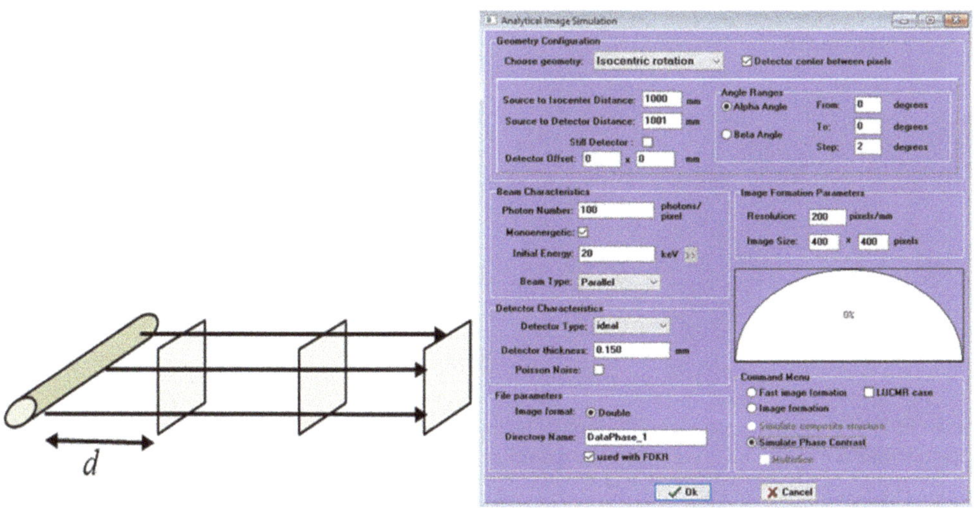

(a) (b)

Figure 7.8. (a) Cylinder object exposed at 20 keV and projection images registered at three different distances, and (b) set-up used to generate phase-contrast images.

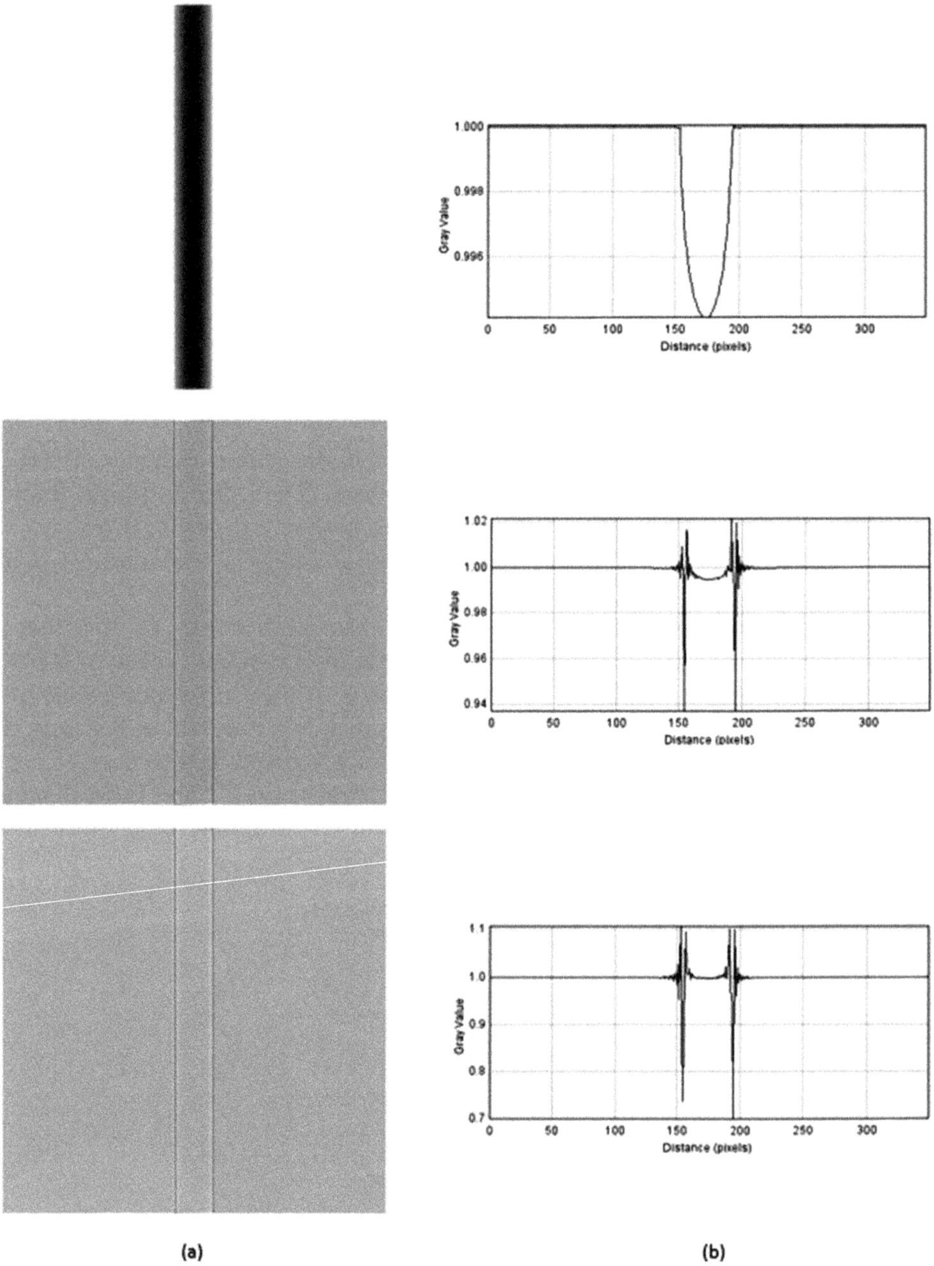

Figure 7.9. (a) Projection images and (b) profiles taken across them for SDD = 1001 mm (first row), SDD = 1300 mm (second row), and SDD = 2300 mm (third row).

edges is well observed for SDDs 1300 mm and 2300 mm. For the source to detector distance of 2300 mm, the phase effects are more pronounced.

7.2.2.3 Virtual study

The aim of this study is to use a computational breast phantom to perform a virtual study to determine if the edge improvement of objects obtained in phase-contrast mode are retained in tomosynthesis mode.

***Methods*:**

- *Involved phantom(s)*: use the *paraffin_phantom.phn* from chapter 4, section 4.3.2, exercise 1.
- *Imaging techniques*: conventional and phase-contrast imaging tomosynthesis. Image resolution: at least 100 pixels/mm; size of images: 2500 pixels × 2500 pixels.
- *Any other parameters* are at your discretion.

Tasks:

(a) Open the *XRAYImagingSimulator* and load the *paraffin_phantom.phn* phantom.
(b) Obtain synthetic conventional mammographic images of the phantom at an energy selected at your discretion (in the range 20 keV).
(c) Obtain phase-contrast mammographic images at a source-to-detector distance (SDD) that guarantees the phase-contrast (~5000 mm). Obtain 21 synthetic phase-contrast images over an arc of −20° to 20° with a step of 2°.
(d) Perform image reconstruction at the central plane of interest.
(e) Compare the images subjectively (visually): planar and tomograms.
(f) Compute objective figures of merit (FOMs)—contrast in the region of the abnormalities and compare.

To be prepared:

- *A report*: a short written summary of the work and the obtained results; containing for example:
  - appropriate outcome images and profiles,
  - tables with computed FOMs,
  - discussion of the results and conclusions.
- *A PowerPoint presentation*: to be used for a short (10 minute) oral presentation.

Useful hints:

- Phase-contrast projection images are generated using the *XRAYImagingSimulator* tool and the option *Simulate Phase Contrast*.
- Image reconstruction is performed using the *FDKR* tool.

The *paraffin_phantom.phn* was loaded in the *XRAYImagingSimulator*. The parameters used to generate 21 phase-contrast images are shown in figure 7.10(a). The generated projection image at 0° is shown in figure 7.11(a). The

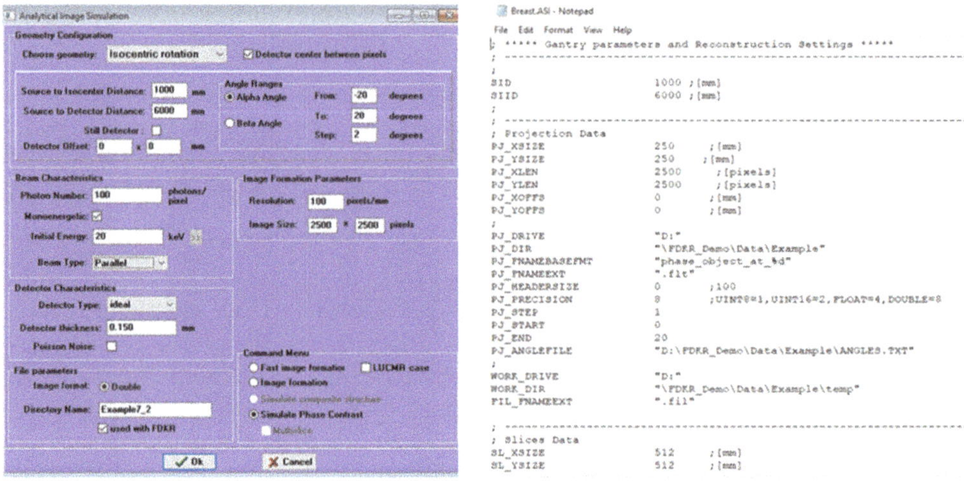

(a) (b)

Figure 7.10. (a) The parameters used to generate a set of 21 phase-contrast images in tomosynthesis mode and (b) the *ASI* file used with the *FDKR* tool to reconstruct the images.

images were then reconstructed with the *FDKR* application with the ASI file shown in figure 7.10(b). A tomogram taken at the isocentre plane is shown in figure 7.11(b).

This is another example of a work project assignment. During the face-to-face session, the participants will create a similar phantom with a different composition and generate radiography and phase-contrast images to be evaluated against clinical ones. An abnormality may be added.

7.2.3 Finding suitable tissue-equivalent materials for physical phantoms for x-ray imaging

The aim of this task is to use computer-based models for preliminary evaluation of different materials as appropriate or not for prototyping of physical phantoms dedicated to x-ray imaging.

Approach:

- Simple phantom configurations are first planned; possible physical implementation is considered.
- Corresponding physical phantoms of different compositions are manufactured (in our case an available CIRS phantom was also added to the study).
- The same phantoms are modelled as computational ones.
- Projection images of these physical phantoms are obtained.
- Projection images of the computational phantoms are generated.
- Images are compared subjectively and by using common figures of merit: line profiles and CNR.

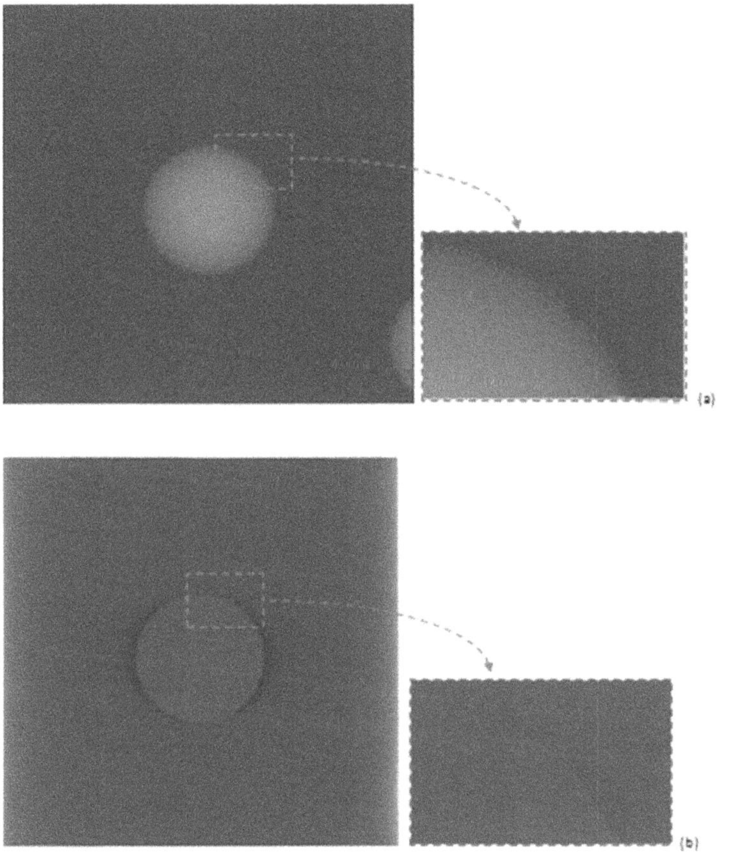

Figure 7.11. Phase-contrast results: (a) a projection image which corresponds to a 0° gantry position and (b) the reconstruction plane which corresponds to the middle plane of the phantom.

References

Arfelli F *et al* 1998 Low-dose phase contrast x-ray medical imaging *Phys. Med. Biol.* **43** 2845–52

Arfelli F *et al* 2000 Mammography with synchrotron radiation: phase-detection techniques *Radiology* **215** 286–93

Asaga T *et al* 1987 Breast imaging: dual energy projection radiography with digital radiography *Radiology* **164** 869–70

Badr S *et al* 2014 Dual-energy contrast-enhanced digital mammography in routine clinical practice in 2013 *Diagn. Interv. Imaging* **95** 245–58

Bliznakova K *et al* 2015 A software platform for phase contrast x-ray breast imaging research *Comput. Biol. Med.* **61** 62–74

Boone J M and Shaber G S 1990 Dual-energy mammography: a detector analysis *Med. Phys.* **17** 665–75

Jacobson B 1953 Dichromatic absorption radiography; dichromography *Acta Radiol.* **39** 437–52

Johns P C *et al* 1983 Dual energy mammographic imaging *Proc. SPIE* **419** 201–11

Johns P C *et al* 1985 Dual energy mammography: Initial experimental results *Med. Phys.* **12** 297–304

Johns P C and Yaffe M J 1985 Theoretical optimization of dual energy x ray imaging with application to mammography *Med. Phys.* **12** 289–96

Lewis R A 2004 Medical phase contrast x-ray imaging: current status and future prospects *Phys. Med. Biol.* **49** 3573–83

Marziani M *et al* 2002 Dual-energy tissue cancellation in mammography with quasi-monochromatic x-rays *Phys. Med. Biol.* **47** 305–13

Taibi A *et al* 2003 Dual-energy imaging in full-field digital mammography: a phantom study *Phys. Med. Biol.* **48** 1945–56

Taibi A and Vecchio S 2014 Breast imaging *Comprehensive Biomedical Physics* 1st edn (Amsterdam: Elsevier) pp 121–54

Chapter 8

Computational models as prototypes of physical anthropomorphic phantoms

8.1 Introduction

The manufacturing of very realistic anthropomorphic phantoms using 3D printing is a challenging task. It is associated with difficulties related to current 3D printing technologies: the limited number of available materials for printing (which in the majority of cases do not appropriately reflect the properties of human tissues), manufacturing precision, limitations in the size of the printed models and their resolution, the format of the software models, etc. Due to these shortcomings, there is a limited number of anthropomorphic physical phantoms for studies in x-ray imaging, either available or currently under development (Winslow *et al* 2009, Carton *et al* 2011, Anderson *et al* 2014, Kiarashi *et al* 2015, Leng *et al* 2015). Thus there is a strong need for further research and technological improvement in this field.

The aim of this chapter is to briefly describe 3D printing, and introduce the relevant terminology and basic 3D printing technologies, as well as to present the use of 3D printers for some biomedical applications and the construction of anthropomorphic phantoms. Two of the 3D printing technologies will be used during practical exercises included in this chapter. The examples of 3D printing applications are selected from the field of x-ray imaging and they are taken from the research work of the Laboratory of Computer Simulations in Medicine at the Technical University of Varna, Bulgaria, the research of which is dedicated to developing, testing and optimising novel x-ray breast imaging techniques.

doi:10.1088/2053-2563/aae197ch8

8.2 Overview of three-dimensional printing

8.2.1 The available printing technologies

3D printing is a prototyping process, in which a real three-dimensional object is created from three-dimensional digital information (a 3D digital model). The 3D digital models are 'sliced' into many two-dimensional cross-sections that are then 'printed' one on top of the other. In other words, the 3D printers create the objects layer-by-layer, and this process is known as desktop fabrication or additive manufacturing. The digital 3D model is saved in a common file format (usually called STL) and then transmitted to the 3D printer for printing (figure 8.1). First you start with a 3D model; you can create the model using relevant 3D software, or scan a real object with a 3D scanner, or download an already made file. Then the 3D printing software 'slices' the model into many 2D cross-sections, thus creating the layers, which the 3D printer then prints one-by-one in a sequence.

Currently, there are three basic and most well-known technologies dedicated to 3D printing: selective laser sintering (SLS), stereolithography (SLA), and fused deposition modelling (FDM). The main differences between the three technologies are in the approach to building the layers, which subsequently results in the properties of the final printed 3D object. There are also other technologies, which arise as a result and combination of the three basic ones and try to exploit the advantages of particular technologies.

In the selective laser sintering technology for fast prototyping (figure 8.2), the 3D printed model is printed from a wide range of powder materials, such as wax, polystyrene, nylon, glass, ceramics, stainless steel, titanium, aluminium, and various alloys. The printed objects are lightweight. The printing process starts with cutting the 3D digital model into thin layers. Each layer is then sent to the 3D printing device. There are two main parts in the SLS technology: a container filled with powder polymer and a device which is used to distribute the powder in the form of thin layers (the roller), as shown in figure 8.2. During the 3D printing process, a fine layer of powder is laid down and then using a laser controlled by the software of the 3D printing device, some of the powder granules are selectively sintered and fused together. These are the granules, the positions of which correspond to the positions in the computer model. When the current layer is printed, the construction platform is lowered by a distance, corresponding to the layer thickness, and a new thin layer of powder is deposited on the printing model and the process is repeated until the prototype is completed. During printing, non-bonded powder granules support the object as it is being constructed. Once printing is completed, almost all of the excess powder can be recycled.

The stereolithography technology produces very small and fine details with perfectly smoothed surfaces, which do not normally require any additional processing. The minimum built layer thickness of this 3D printer may be as thin as 0.02 mm.

Figure 8.1. Schematic flow of the 3D printing process.

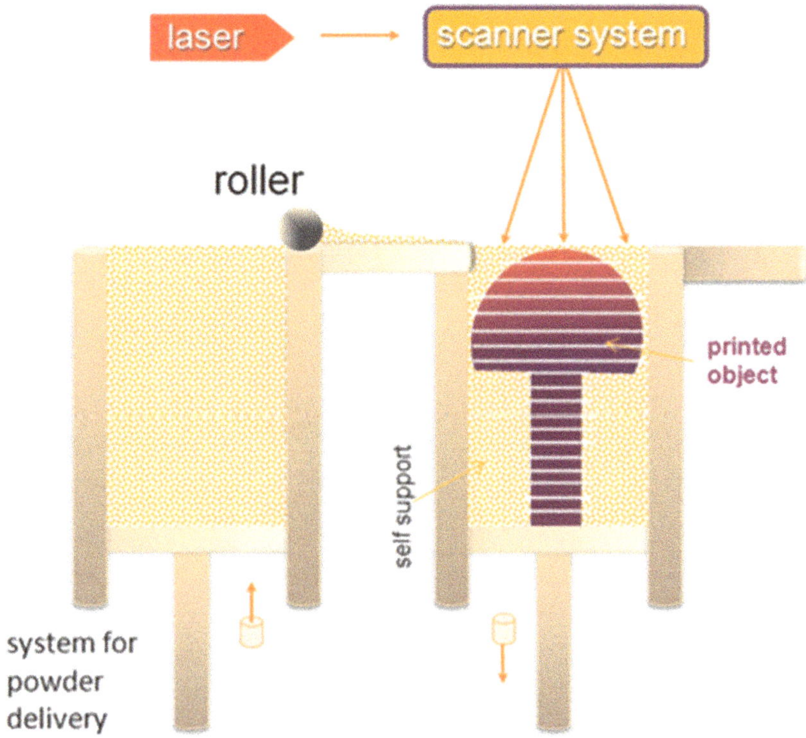

Figure 8.2. Schematic presentation of the selective laser sintering (SLS) 3D printing technology.

Stereolithography 3D printers position a moving base/platform just below the surface of a vat of liquid photocurable polymer/resin (figure 8.3). An ultraviolet laser beam then traces the first layer of the printed object on the surface of the liquid polymer, causing a very thin layer of photo-induced polymer to harden. The base is then lowered or raised slightly, (depending on the type of printer), which corresponds to the layer thickness. Then, another layer is traced out, hardened by the laser, and produced. This process continues until the complete object is built up. Then, the object is removed from the vat of photopolymer, drained of excess liquid, and cleaned. The SLA printers remain among the most accurate for the fabrication of 3D objects.

In the fused deposition modelling technology, the 3D printed object is created layer by layer through heating the thermoplastic material to a semi-liquid state, and extrusion and deposition of the material, according to a computer model (figure 8.4). FDM may use more than one material to accomplish the process: e.g. one material for the printed object itself, and another auxiliary material, either for the printed object again (to create objects of two or more different materials), or a material that is soluble and may be used as a base or a supporting structure. Objects are produced with a high degree of accuracy. Most FDM 3D printers can print with acrylonitrile butadiene styrene (ABS), biodegradable bioplastic called PLA–polylactic acid, which is produced from organic alternatives of oil. A key benefit of FDM printers is that objects can be made of out of the same thermoplastics used in traditional injection moulding.

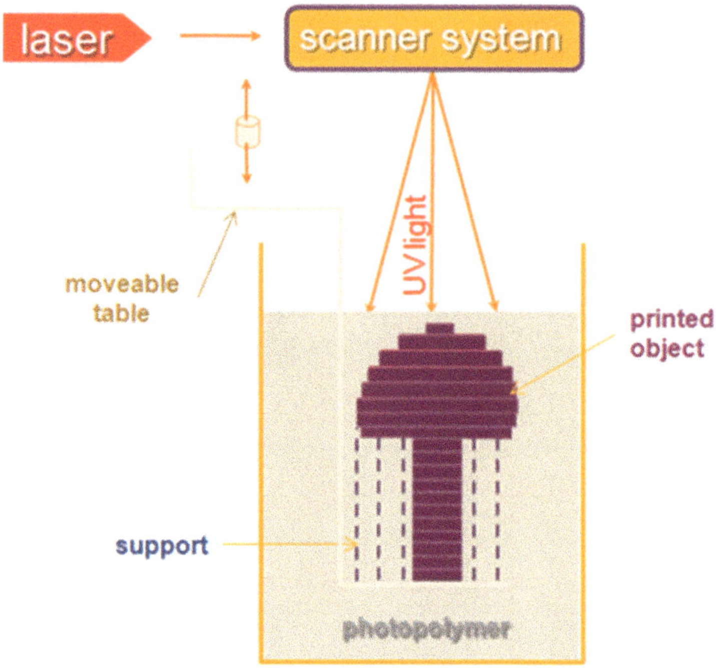

Figure 8.3. Schematic presentation of the stereolithography (SLA) 3D printing technology.

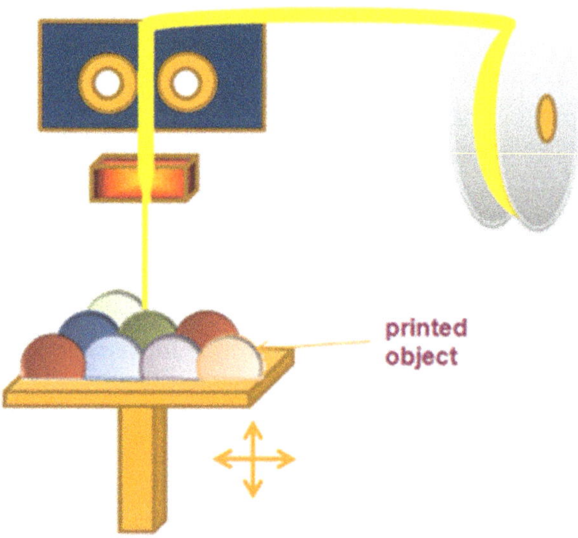

Figure 8.4. Schematic presentation of the fused deposition modelling (FDM) 3D printing technology.

Another 3D printing technology, which is a combination of FDM and SLA technologies, is called polyjet matrix. This technology allows simultaneous jetting of different photo-curable polymer materials—over ten at a time and in certain models up to 30 from the print head, which are then hardened by a laser beam. The key benefits of this technology is that it creates multiple-material printed objects with a high accuracy and a high speed of printing.

In summary, the printing resolution of typical current 3D printers is between 300 DPI (along *x–y–z*) and 900 DPI (along *x–y–z* in ultra-HD resolution), while the corresponding voxel size ranges between 0.025 mm and 0.050 mm. The dimensions of the printed objects may typically reach values of up to 1000 mm along the *x*, *y*, *z* directions, and up to 1500 mm along a single direction only.

Currently, more and more materials suitable for 3D printing are appearing on the market. They can be classified as:

- *Metallic materials*: plain carbon steel, tool steel, stainless steel, aluminium, copper, titanium, bronze, nickel alumides.
- *Polymers and polymeric composites*: ABS, nylon (polyamide), polycarbonate, polypropylene, epoxies, glass filled polyamide, windform, polystyrene, polyester, polyphenylesulfone.
- *Others*: sand, ceramics, elastomers, tungsten, wax, starch, plaster.
- *Bio-compatible materials*: polycaprolactone (PCL), polypropylene-tricalcium phosphate, (PP-TCP), PCL-hydroxyapatite (HA), polyetheretherketone-hydroxyapatite, (PEEK-HA), tetracalcium phosphate (TTCP), beta-tricalcium phosphate (TCP), polymethyl methacrylate (PMMA).

However, 3D printing technologies still suffer from several limitations, such as the cost, accuracy, finish of the printed surfaces, strength of the final product, available printing materials, etc. These aspects also reflect the main criteria for choosing a specific 3D printing technology: the cost of the 3D printer, cost of the printed prototype, printing speed, choice of materials, colour capabilities, etc. Table 8.1 shows a schematic comparison between the three main 3D printing technologies.

8.2.2 Examples of printed phantoms—aspects for consideration

For the purposes of the EUTEMPE Anthropomorphic Phantoms course, the Laboratory of Computer Simulations in Medicine at the Technical University of Varna uses two 3D printers: the first one is based on the FDM technology (LeapFrog Creatr 3D dual head printer, www.lpfrg.com) and the second one is

Table 8.1. Schematic comparison between the three main 3D printing technologies.

	Accuracy	Cost	Advantages	Disadvantages
SLS	++	$$$	Large part size, variety of materials, good strength	High cost, powdery surface
SLA	+++	$$	High accuracy, high speed	Moderate strength, small part size
FDM	++	$	Low cost, good strength	Low speed

based on the SLA technology (FormLabs FORM 1+, https://formlabs.com/). These printers are used daily in preparing different phantoms dedicated to research and educational purposes in the field of x-ray imaging.

These printers were tested to produce two test objects of particular interest for the research work of the team. One of the objects was a compressed breast phantom, shown in figures 8.5(a), (b), and the other one was a dental construction shown in figures 8.5(c), (d). In figures 8.5(a), (c), the objects are printed with the SLA printer, while in figures 8.5(b), (d) the objects are printed with the FDM printer. As seen from these images, the objects printed with the SLA technology are characterized by smooth surfaces compared to the objects printed with the FDM printer. Both models are printed with a 0.1 mm layer thickness. Moreover, the samples from the SLA printer are printed using clear resin material. The objects in figures 8.5(a), (b) are used in studies of x-ray breast imaging techniques, while the dental constructions (figure 8.5(c), (d)) are used in a collaborative study with the Faculty of Dental Medicine, Medical University of Varna, Bulgaria.

Speed and temperature are key factors in the FDM 3D printing process. Failing to correctly choose one of these parameters may result in defective printed objects, as demonstrated in figure 8.6.

For the test object printed in figure 8.6(a), the defective printing was due to both unsuitable speed and temperature being selected for the specific material. Figure 8.6(b) shows a trial for printing a duct tree of the breast phantom using two ABS materials simultaneously, using inappropriate speed settings. Defective printing with the SLA printer is demonstrated in figure 8.6(c). The reason for this defect is still under clarification.

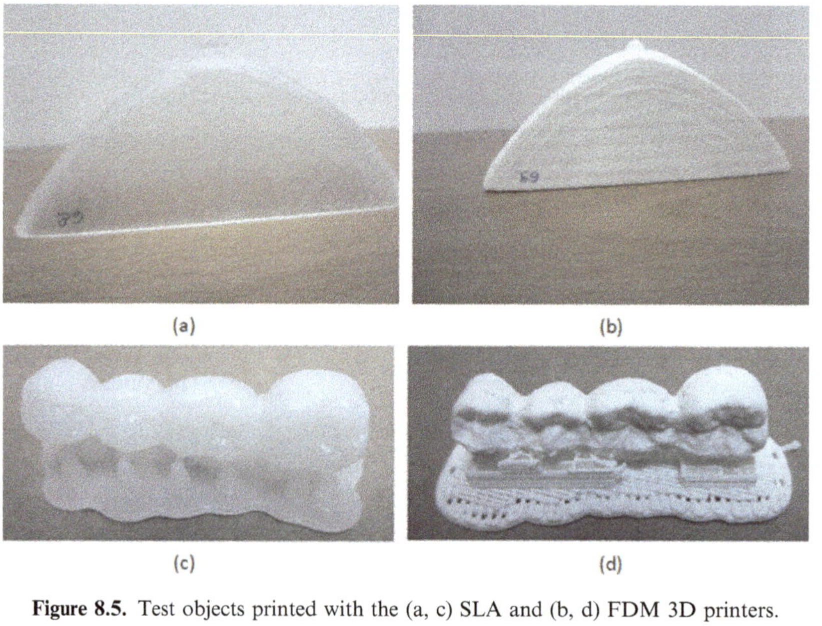

Figure 8.5. Test objects printed with the (a, c) SLA and (b, d) FDM 3D printers.

(a) (b) (c)

Figure 8.6. Demonstration of defective print-outs due to incorrect printing process settings.

During the last few years, the application of 3D printing in biomedical engineering research has grown enormously. A basic requirement is the digital (software) model. Such models may be obtained from either dedicated CAD software that can output computational 3D digital models, or from 3D imaging techniques such as MRI, CT, PET, etc, that can provide 3D data by means of voxel matrices.

An example of the first option is described for the specific case of breast phantoms. A computational model of the breast is designed with an in-house developed software (Bliznakova *et al* 2010, 2012), dedicated to generation of complete computer-based realistic breast models for x-ray imaging and their use in virtual experiments that aim at design, testing, and optimizing of new x-ray breast imaging techniques (figure 8.7(a)). Then, the computer model is 'passed' onto the 3D printer in order to create the physical model (figure 8.7(b)).

The second possibility is to create 3D computational objects from images acquired using 3D diagnostic medical imaging techniques. One such example is demonstrated in figure 8.8 for the case of the liver.

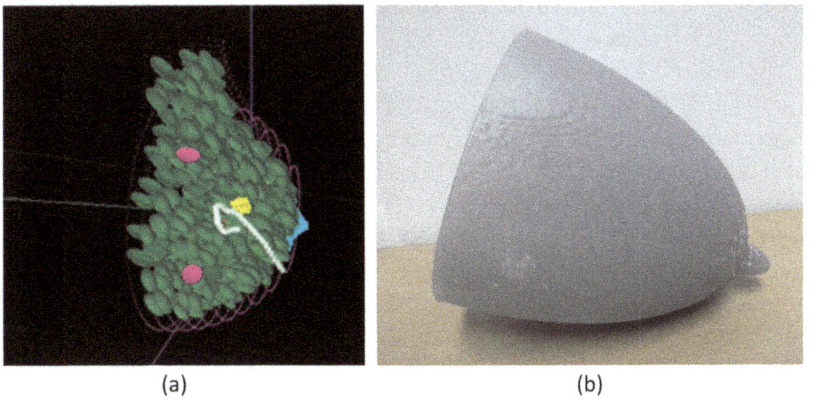

(a) (b)

Figure 8.7. Breast phantom developed at the Technical University of Varna: (a) the computer model and (b) the physical phantom.

Figure 8.8. Creating computational objects from patient-based CT images: (a) a CT image extracted from a patient scan, (b) the segmented object of interest—the liver, and (c) the 3D printed physical phantom of the liver.

The liver shape is automatically selected from patient abdominal CT images and thus the region that belongs to the liver was segmented from the other tissues and used in preparing the final 3D liver volume.

8.3 Printing 3D prototypes from computational objects

8.3.1 Pre-processing of computational objects prior to 3D printing

Before the actual 3D printing process takes place, several preparatory steps need to be accomplished. These include pre-processing of the computational objects/models, preparation of the file for printing (the *.stl file), and importing the file into the printing software.

The first step is using relevant CAD software to create the 3D digital model. Then, the model is subjected to three-dimensional processing, mainly including smoothing of the surfaces in order to achieve better print quality. Afterwards, the 3D digital model is exported into STL format and the *.stl file is then passed onto the 3D printer to create the physical model (figure 8.9).

STL is a file format native to the stereolithography CAD software created by 3D Systems (3D Systems Inc 1989). STL has several after-the-fact backronyms such as 'Standard Triangle Language' and 'Standard Tessellation Language' (Grimm 2004). This file format is supported by many other software packages; it is widely used for rapid prototyping, 3D printing, and computer-aided manufacturing (Noorani 2006). STL files describe only the surface geometry of a three-dimensional object without any representation of colour, texture, or other common CAD model attributes (figure 8.10). The STL format specifies both ASCII and binary representations. Binary files are more common, since they are more compact.

An STL file describes a raw, unstructured triangulated surface by the unit normal and vertices (ordered by the right-hand rule) of the triangles using a three-dimensional Cartesian coordinate system. In the original specification, all STL coordinates were required to be positive numbers, but this restriction is no longer enforced and

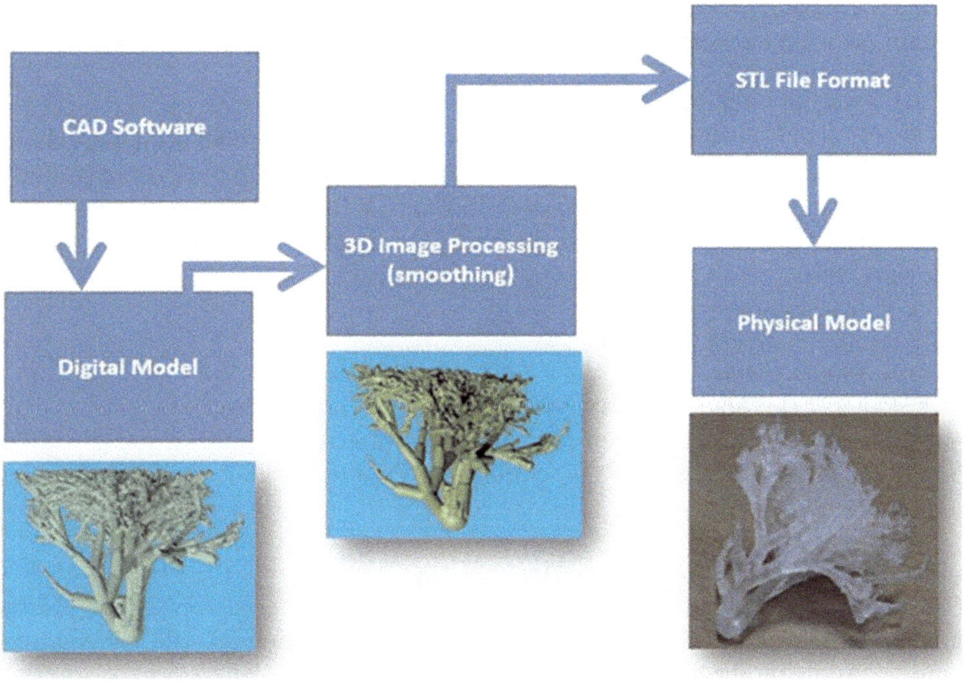

Figure 8.9. Schematic representation of the overall procedure.

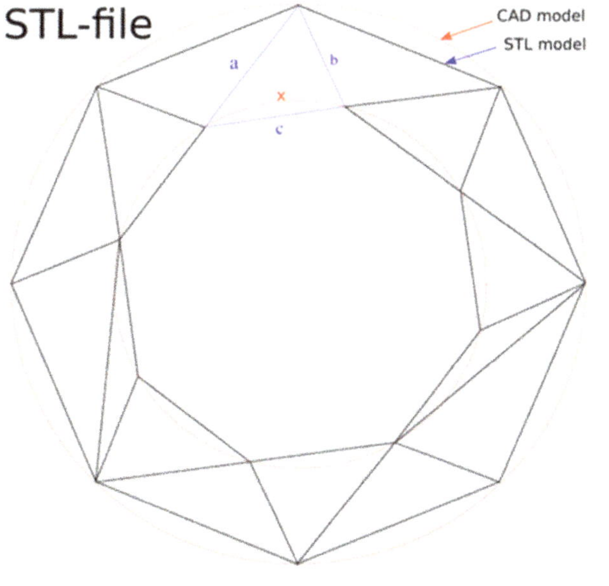

Figure 8.10. Differences between CAD and STL model representations (https://commons.wikimedia.org/wiki/File:The_differences_between_CAD_and_STL_Models.svg).

negative coordinates are commonly encountered in STL files today. STL files contain no scale information, and the units are arbitrary.

8.3.1.1 Open the 3D digital model file

The first step in the overall process is to open the 3D digital model using the relevant software package. In our exercises the *MATLAB* (https://mathworks.com/products/matlab.html) or *Octave* (https://octave-online.net/) software packages will be used. A three-dimensional volume containing the 3D model can be opened using the following script:

Nx = 500; % *pixels*
Ny = 500; % *pixels*
Nz = 500; % *pixels*
F = *rawread3d*('*breast.bin*',*Nx*,*Ny*,*Nz*,'*single*');

where Nx, Ny, Nz are the dimensions of the 3D matrix along the x, y, z directions, '*breast.bin*' is the name of the file, containing the model, and '*single*' is the single precision (32-bit float).

8.3.1.2 Segmentation of the material

The next step is to 'convert' the 3D model into two materials only. The printing material will be assigned to a gland, and another filling material will be assigned to adipose. Applying the script below:

$$F((F > 0.07)\&(F < 0.08)) = 100;$$
$$F(F\tilde{} = 100) = 220;$$

reduces the values of the 3D model (3D matrix) to two numbers (100 and 220) only, thus converting the model to two materials only.

In the case of large volumes, there may be a need to reduce the 3D matrix, using the following script:

$$Fr = F(1{:}\ 2{:}\ Nx,\ 1{:}\ 2{:}\ Ny,\ 1{:}\ 2{:}\ Nz).$$

8.3.1.3 Smoothing

Smoothing is used to reduce the effects from the voxelisation of the 3D model. The 3D models/objects are presented in the computational domain as composed of a multitude of tiny cubes, so-called voxels. In contrast to the originally smooth object, the voxel model has a jagged surface. The stresses induced by the computed displacements can have singularities at the edges and corners of the surface, but also at interfaces between different materials. A straightforward approach to get rid of the singular stresses is to smooth the surface and material interfaces of the computational domain (figure 8.11).

The most basic of filtering operations is called 'low-pass'. A low-pass filter, also called a 'blurring' or 'smoothing' filter, averages out rapid changes in intensity. The simplest low-pass filter just calculates the average of a pixel and all of its eight immediate neighbours. The result replaces the original value of the pixel. The process is repeated for every pixel in the image.

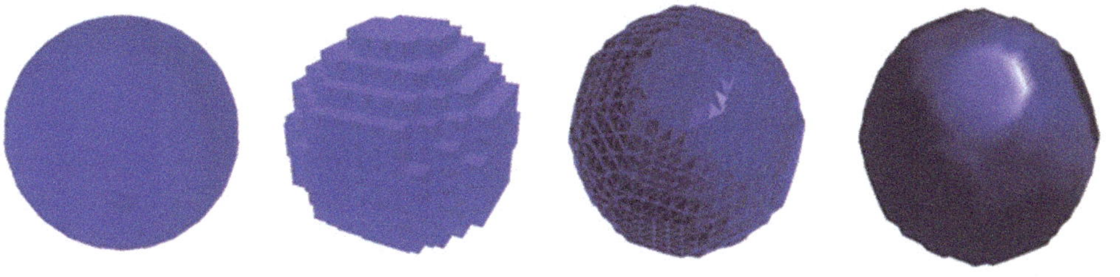

Figure 8.11. Visual presentation of the smoothing process.

In *MATLAB*, you can use the *smooth3* function:

W = *smooth*3(*V*), smooths the input data *V* and returns the smoothed data in *W*.

W = *smooth*3(*V*, *'filter'*), the *'filter'* determines the convolution kernel and can be the strings *'gaussian'* or *'box'* (default).

W = *smooth*3(*V*, *'filter'*, *size*), sets the size of the convolution kernel (the default is [3, 3, 3]). If *size* is scalar, then *size* is interpreted as [size, size, size].

W = *smooth*3(*V*, *'filter'*, *size*, *sd*), sets an attribute of the convolution kernel. When the filter is *gaussian*, *sd* is the standard deviation (the default is 0.65).

*8.3.1.4 Preparation of the *.STL file*

In order to prepare the *.STL file, first we need to extract the 'isosurface' data from the 'volume' data. For this purpose, the following scripts are applied:

[*Nx*, *Ny*, *Nz*] = *size*(*Fr*);

[*x*, *y*, *z*] = *meshgrid*(1:*Nx*, 1:*Ny*, 1:*Nz*);

fv = *isosurface*(*x*, *y*, *z*, *Fr*, 195);

Then, the data can be saved in a *.STL file using the following function:

stlwrite(*'breast.stl'*, *fv*);

where *'breast.stl'* is the name of the output file and *fv* is the data matrix. The created *.STL file can be visualized in *MATLAB*/*Octave* using the following function:

patch(*fv*);

The created *.STL file can also be imported and visualized in a dedicated 3D printer software package.

8.3.2 Tutorial for 3D printing

In this tutorial, we will have a look at two of the common 3D printing technologies already described above—the FDM and the SLA.

8.3.2.1 FDM 3D printing method—the Leapfrog printer

Create *.STL file or *.OBJ file

In order to begin 3D printing, first step is creating a model of the desired object. Most 3D printers' software accept *.STL and *.OBJ files and convert them to g-code file (executable code). The model file can be downloaded or created by a CAD system (*AUTOCAD*, *Sketch up*, *Designspark mechanical*, *Solidworks*, and others). In this course the model will be designed using *Breast Simulator* software and image processing will be accomplished under the *MATLAB/Octave* software environment.

***Import the STL file in Simplify3D* software**

Importing a file in *Simplify3D* is very easy. Click over *Import* button (figure 8.12(a)) and browse to the desired file. The drag and drop function is also available. Once imported, the file is shown in the *Models* list (figure 8.12(b)).

Model settings

Double-clicking on the model opens the model settings window (figure 8.13) that gives you the opportunity to scale, move, or rotate your object. The size of the object can be also observed. As you make changes, you will see the impact of each adjustment as your model reacts and moves on the *Build* table. Click *Save* to apply those changes to the selected model(s). Note that you can also change the display name of your model through this window.

Here are some general mouse shortcuts:

Zoom—scrolling (or Shift-left-click and drag up and down).

Rotate View—left-click and drag.

Pan View—right-click and drag.

Move Model—Ctrl-left-click and drag.

Scale Model—Ctrl-right-click and drag up or down.

Rotate Model—Alt-left-click and drag right or left.

The *Center and Arrange* button (figure 8.12(b)) will centre your model automatically. It is a very helpful tool when your model is far outside the build table. For

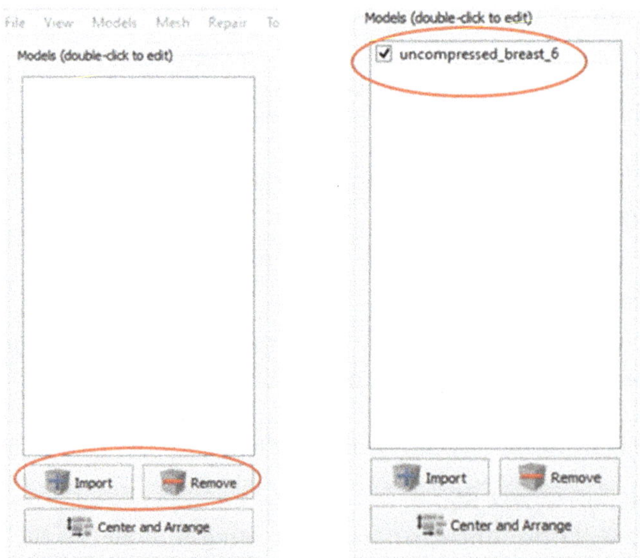

(a) (b)

Figure 8.12. Import STL file: (a) import/remove file and (b) the inserted file.

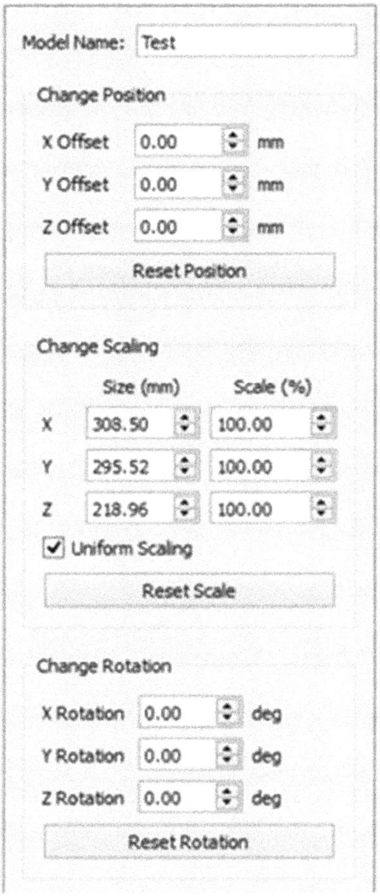

Figure 8.13. Model settings window.

more information about importing and manipulating your models visit https://www.simplify3d.com/support/tutorials/importing-and-manipulating-your-models/.

Add process

The 'process' is a collection of settings including speed, size, temperature, support material, quality levels, etc. Click the *Add* button in order to create a new process (figure 8.14(a)). *3DSimplify* will add the last used process. To make adjustments to your process, double-click on the process or click *Edit Process Settings* (figure 8.14(b)).

This will open the *FFF Settings* window (figure 8.15), where you can make common settings for your model.

Process name

You can leave this section unchanged. But in the case of using two or more processes, it is good practice to give them names.

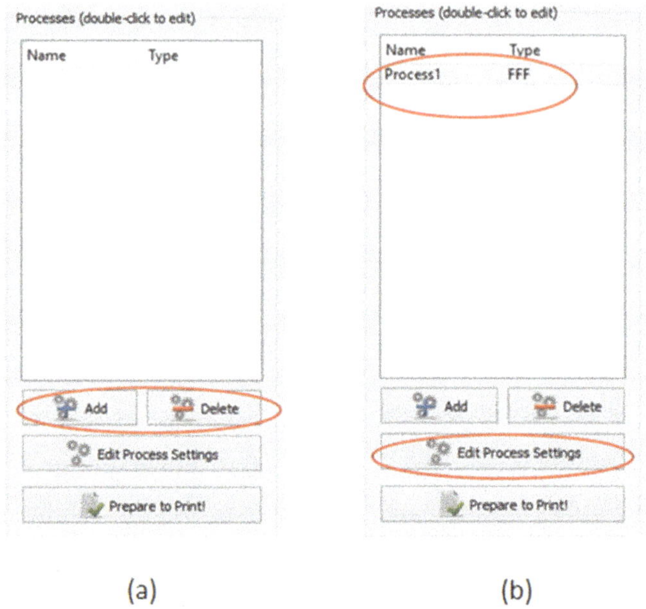

Figure 8.14. Add and edit a process: (a) add/delete process and (b) edit process.

Select 3D profile
There are different profiles according to the type of printer and the number of extruders. Every profile has predefined parameters that users can modify for the actual print model. Saving your own profile is also nice option. The printer is Leapfrog Creatr Dual. Choose the left, right, or both extruders option according to the print model.

Select material
The Leapfrog 3D Printer can work with different materials such as ABS, PLA, PVA, and Nylon. In our exercises we will use two of them: ABS and PLA. According to the printed model it is possible to print the whole object with one material or a combination of both. Choose the desired material. You can read more about printing with two materials at http://www.simplify3d.com/support/tutorials/.

Print quality
Different types of print quality mainly mean that layer thickness will be different, which results in different speeds of printing. Medium quality (0.2 mm) is a good choice for basic prints.

Raft
A raft is a horizontal latticework of filament that is located underneath your part. rafts are primarily used with ABS to help with bed adhesion. Rafts are also used to help stabilize models with small footprints, or to create a strong foundation on which to build the upper layers of your part. Chose the *Raft* option in the *Process*

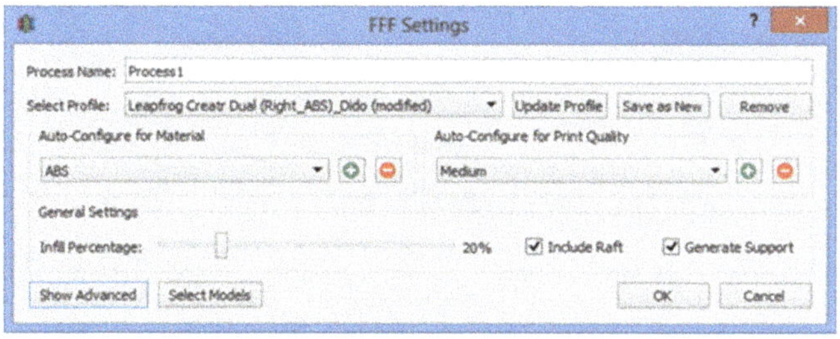

Figure 8.15. *FFF Settings* window.

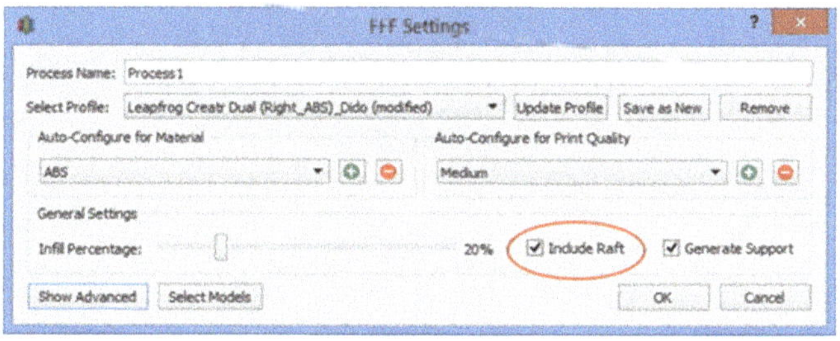

Figure 8.16. *Raft* option.

Settings window. For more information visit https://www.simplify3d.com/support/tutorials/rafts-skirts-and-brims/.

Support

Support material is added during the printing process in order to support the overhanging parts of your design. Not all shapes need support material. If there are overhang parts with more than 45 degrees angles, support structures have to be added to the design. Clicking the *Support* button (figure 8.16) opens the *Support Generation* window (figure 8.16), which allows the user to add or delete support structures manually. For more detailed information visit https://www.simplify3d.com/support/tutorials/adding-and-modifying-support-structures/.

Infill

Infill changes the interior solidity of your 3D print. 0% is completely hollow and 100% is completely solid. A value in the range of 20%–50% is a good place to start.

Advanced options

Advanced settings can be found by clicking *Show Advanced* (figure 8.17(a),(b)) at the bottom left corner of the *FFF Settings* window. This reveals all of the detailed settings that the *Simplify3D* software is implementing behind the scenes, as well as providing customization options for many settings you may want to adjust. In this exercise we will change only a few basic option such as layer thickness, temperature of extruders and bed, and speed. Be careful as incorrect adjustments may lead to faulty printed models.

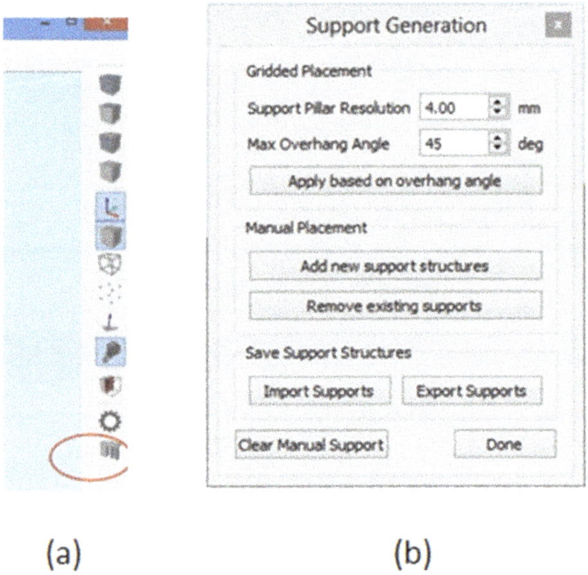

Figure 8.17. Support generation: (a) the *Support* button and (b) the *Support Generation* window.

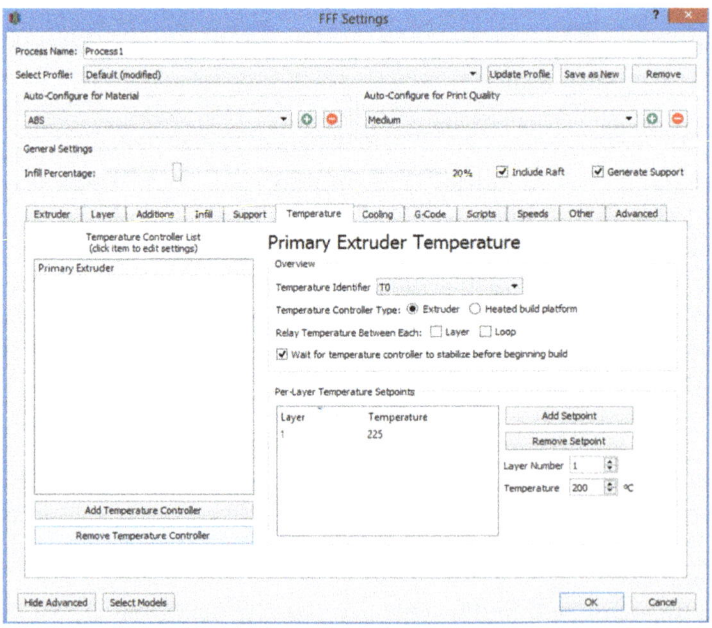

Figure 8.18. Temperature adjustment.

For temperature adjustment (figure 8.18), temperatures were empirically tested for different materials. For ABS we advise to start with a bed temperature of 55 °C and an extruder temperature of 230 °C. For PLA we advise to start with a bed temperature of 25 °C and extruder temperature of 210 °C. The suitable temperature

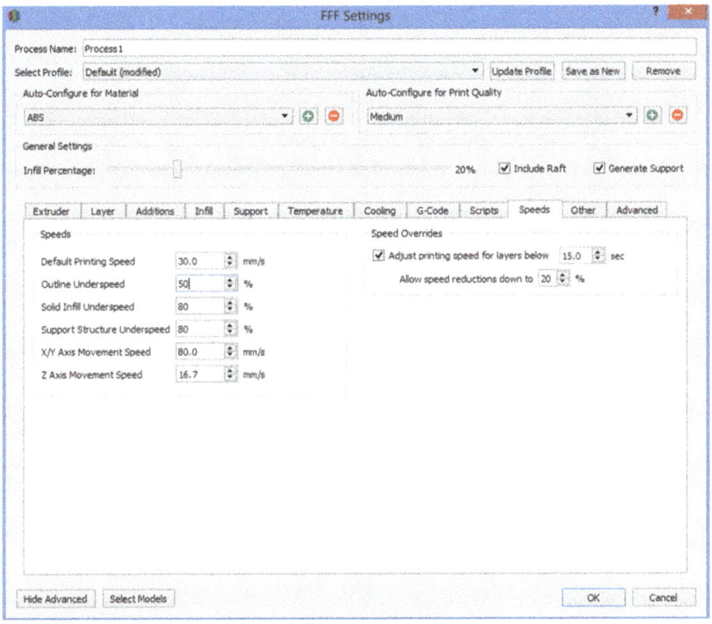

Figure 8.19. Speed adjustments.

may be different for every printed model. Factors that affect this are: ambient temperature, speed, material, structure of objects, etc.

For speed adjustments (figure 8.19), speeds were empirically tested for different materials. For ABS we advise to start with a printing speed of 40–50 mm s^{-1}. For PLA we advise to start with a printing speed of 20–30 mm s^{-1}.

Prepare to print

When all settings are chosen, click the *Prepare to print* button in the bottom left corner. The software will convert your model with its features to a g-code file and the software window will transition automatically to 'preview mode'. The animated preview mode in the *Simpify3D*™ software allows you to inspect all aspects of your print operation prior to execution. The software takes your generated g-code and virtually displays line by line how your model will be constructed (figure 8.20). Take your time to check if your model will be printed as expected. If any corrections are necessary, exit the preview mode and make the necessary changes.

Save the g-code file, factory file

When you are satisfied with the preview, there are two options to begin your print:

Begin printing over USB: if you are using a USB connection, you have the option to view the build sequence in real-time on your monitor by selecting *Live Preview Tracking*. For additional information on USB communication, access the *Machine Control Panel* to view the broad range of information that is transmitted via USB.

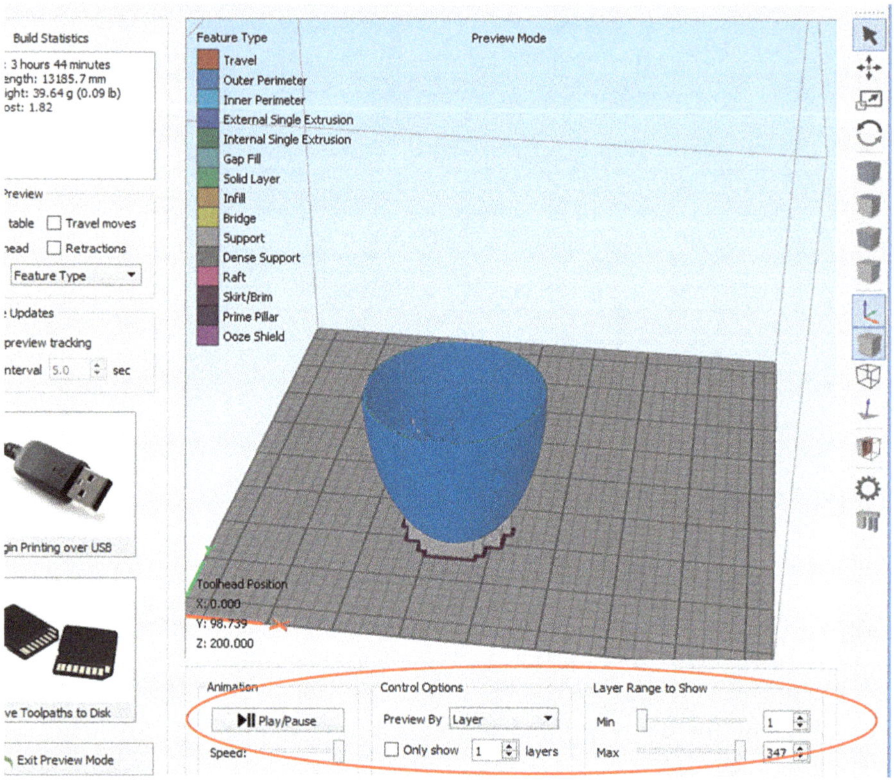

Figure 8.20. Preview mode.

Save toolpaths to disk: this saves your print file to a user-defined location, such as a folder on your computer or an SD card. Then printing will be initiated from the *Machine Control Panel.*

Start printing

First, make sure the print bed is heated to the correct temperature for the filament that you will be using. It is advised to extrude a bit of filament before you run your print job to make sure that the nozzle is not clogged. Please note that the print bed will automatically move to the *Home* position when starting a new print. The home position is the position where the value for the *X*-, *Y*-, and *Z*-direction is zero. You can see that the bed is placed at the highest point and the extruders are positioned at the front left side of the print bed. If you want to print a pre-existing g-code you need to go to the *Machine Control Panel* (figure 8.21(a)) and click *Print* which can be found next to the *Connect/Disconnect* button. A menu will pop-up where you can browse your computer for the g-code you wish to print. Double click the desired g-code and the print job will start.

The extruder carriage or bed may not move immediately after reaching the home position. The printer will stabilize the extruder and bed temperature before it starts any movements. You can monitor this process in the *Communication* tab.

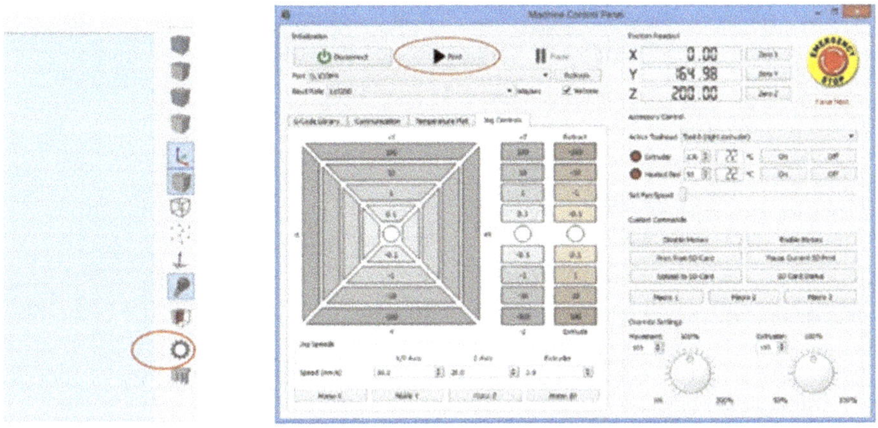

(a) (b)

Figure 8.21. Machine control panel: (a) open panel button and (b) the panel window.

8.3.2.2 Stereolithography method (Formlabs printer)

Create *.STL file or *.OBJ file

In order to begin 3D printing, the first step is creating a model of the desired object. Most 3D printers' software accept *.STL and *.OBJ files and convert them to g-code file. The model file can be downloaded or created by a CAD system (*AUTOCAD*, *Sketch up*, *Designspark mechanical*, *Solidworks*, and others). In this course the model will be designed by *Breast Simulator* software (Bliznakova *et al* 2012) and image processing will be accomplished using *MATLAB/Octave*. Use the model created in the previous section.

Printer set-up

When the PreForm software is started, the *Printer Setup* window appears (figure 8.22). Set the layer thickness and material that will be used. A layer thickness of 0.1 or 0.2 mm is a good starting point for the first print. Select the necessary material for your prints. This window can be re-called again at any time from the bottom tab of the screen (figure 8.23).

Import *.STL file in the PreForm software

Click *File* → *Open* and browse to the desired file (figure 8.24).

Model settings and orientation

Scale the object—adjust the desired size of the model

Click the *Size* button and the *Size* window will appear (figure 8.25). Left-click over the model and drag in order to size the object. For more precise scaling, set coordinates directly in the *Size* window.

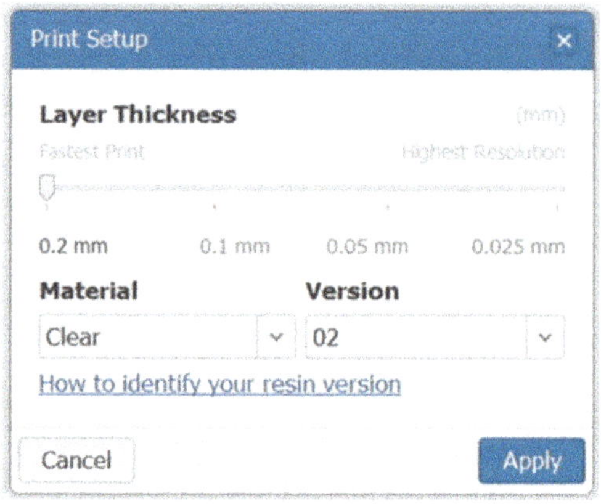

Figure 8.22. Printer set-up.

Figure 8.23. Change settings.

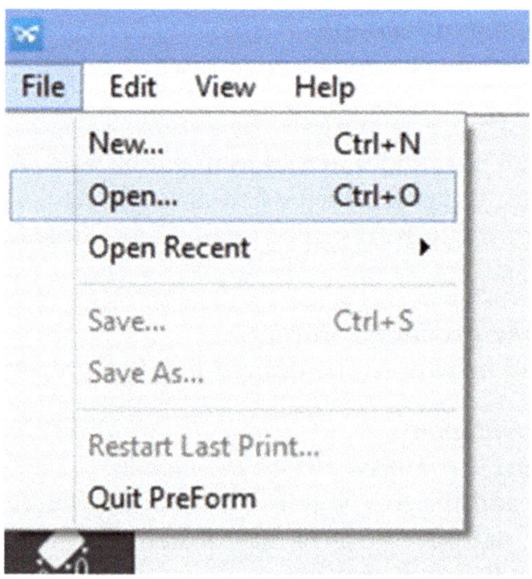

Figure 8.24. Import file.

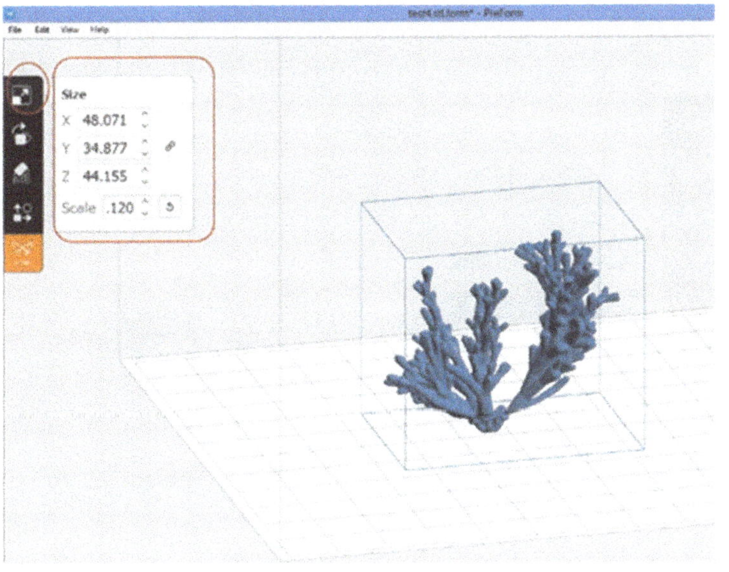

Figure 8.25. The *Size* window.

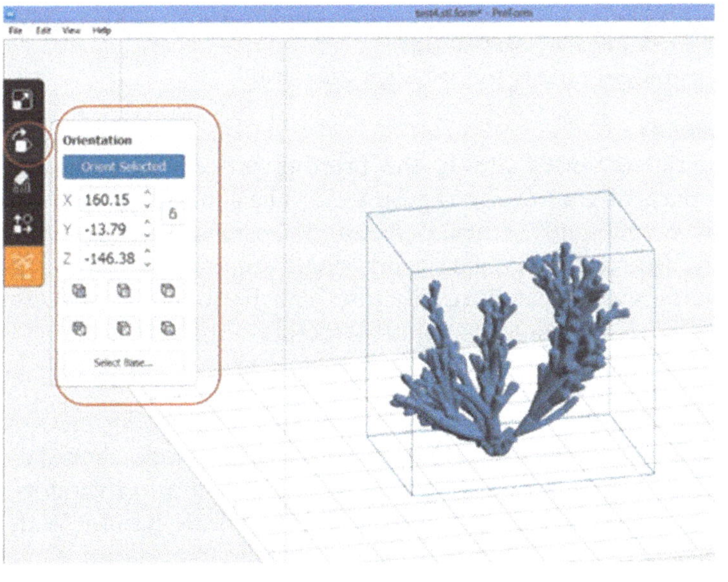

Figure 8.26. The *Position* window.

Position object

Click the *Orientation* button and the *Orientation* window will appear (figure 8.26). Left-click over the model and drag in order to position the object. For a more precise position, set the coordinates directly in the orientation window.

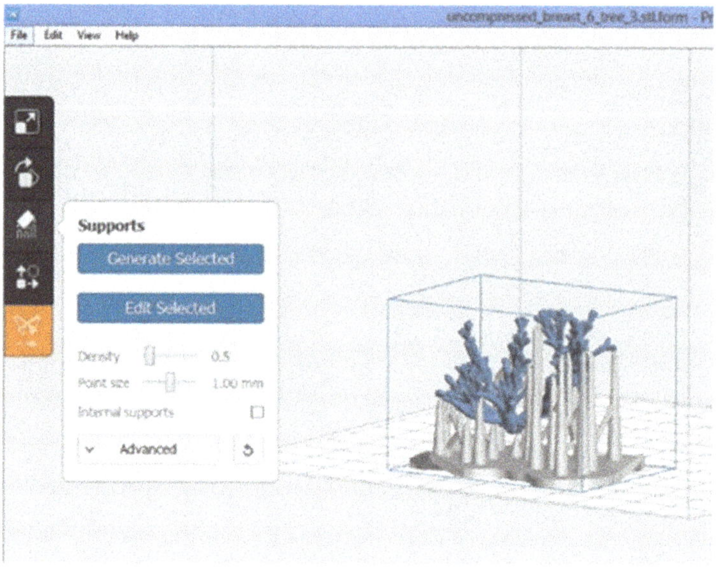

Figure 8.27. Support structures.

General shortcuts:
Zoom—scrolling.
Rotate View—right-click and drag.
Pan View—middle mouse button click and drag.

Support structures

Support material is added during the printing process in order to support the overhanging parts of your design (figure 8.27). The support structures can be added automatically or manually. Click *Generate Selected* to set the support structures automatically. Manually changing every single support structure is also possible. More information about support structures can be found at http://formlabs.com/support/guide/prepare/advanced-support-structure-settings/.

Start printing

Click the printer button once in order to initiate the printing process. On screen, *ready to print* will be displayed. When all settings are made click the *Print* button (figure 8.28) and at the bottom of the screen a configuration window will appear. Check all settings once again and click the *Send to printer* button. Wait until the file is uploaded and click the printer button once more to confirm printing. Clicking the button once during the printing process will pause the printing. Printing is resumed when you click the button again. For cancellation, hold the button for a few seconds.

After printing

When the printing process is finished, the object has to be cleaned in isopropyl alcohol (IPA). More detailed information about post-printing action can be obtained at http://formlabs.com/support/guide/finish/post-print-steps/.

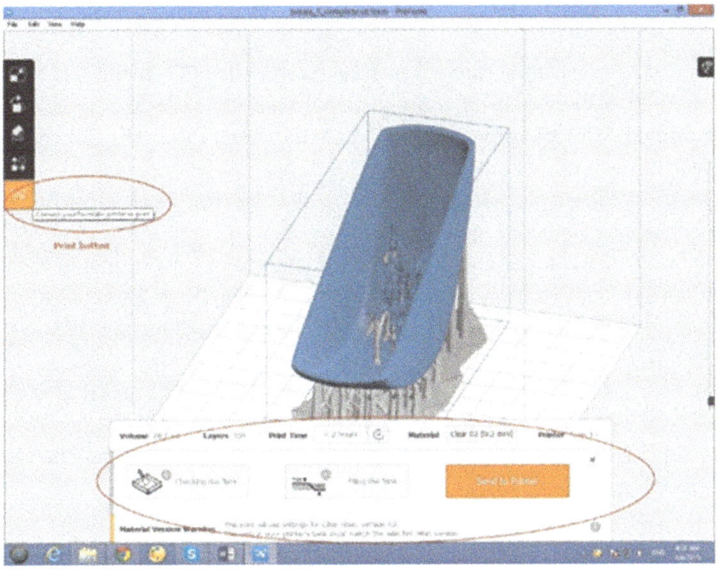

Figure 8.28. Printing window.

8.4 Examples of using 3D printing

8.4.1 Design and manufacture of a simple breast duct tree

A self-study video tutorial is available at the following link:
http://tuv.eutempe-net.eu/wp-content/uploads/Education-materials/Videos/Printing%20Simple%20Breast%20Tree-1%20optimized.mp4.

8.4.2 Design and manufacture of a complex breast duct tree

A self-study video tutorial is available at the following link:
http://tuv.eutempe-net.eu/wp-content/uploads/Education-materials/Videos/Printing%20Complex%20Breast%20Tree-1%20optimized.mp4.

8.4.3 Design and manufacture of a complete breast model

A self-study video tutorial is available at the following link:
http://tuv.eutempe-net.eu/wp-content/uploads/Education-materials/Videos/Printing%20Complete%20Breast-1%20optimized.mp4.

8.5 Test questions

(a) List several advantages and disadvantages of selective laser sintering 3D printing technology.
(b) Why is pre-processing of the computer models necessary before actual 3D printing?
(c) Which 3D printing technology allows printing with more than one material at the same time?
(d) Which 3D printing technology requires support structures?

References

3D Systems, Inc. 1989 *StereoLithography Interface Specification* (Valencia, CA: 3D Systems, Inc.)

Anderson J R *et al* 2014 Validation of computational fluid dynamics methods with anatomically exact, 3D printed MRI phantoms and 4D pcMRI *Conf. Proc. IEEE Eng. Med. Biol. Soc.* **2014** 6699–701

Bliznakova K *et al* 2012 BreastSimulator: a software platform for breast x-ray imaging research *J. Biomed. Graphics Comput.* **2** 1–14

Bliznakova K *et al* 2010 Evaluation of an improved algorithm for producing realistic 3D breast software phantoms: application for mammography *Med. Phys.* **37** 5604–17

Carton A K *et al* 2011 Development of a physical 3D anthropomorphic breast phantom *Med. Phys.* **38** 891–6

Grimm T 2004 *User's Guide to Rapid Prototyping* (Southfield, MI: Society of Manufacturing Engineers)

Kiarashi N *et al* 2015 Development of realistic physical breast phantoms matched to virtual breast phantoms based on human subject data *Med. Phys.* **42** 4116–26

Leng S *et al* 2015 Construction of realistic liver phantoms from patient images using 3D printer and its application in CT image quality assessment *Proc. SPIE* **9412** 94124E

Noorani R 2006 *Rapid Prototyping: Principles and Applications* (New York: Wiley)

Winslow J F *et al* 2009 Construction of anthropomorphic phantoms for use in dosimetry studies *J. Appl. Clin. Med. Phys.* **10** 2986

Chapter 9

EPILOGUE

Dear Reader,

Since you reached this point, we, the authors, believe that you have enjoyed reading this book, dedicated to the module on Anthropomorphic phantoms for optimisation of dose and image quality in radiology from the European Network for Training and Education of Medical Physics Experts—EUTEMPE (www.eutempe-net.eu).

We hope that you have broadened your knowledge in medical physics and biomedical engineering areas and that you have developed new specific skills in the design, creation, and use of anthropomorphic phantoms in virtual clinical trials. Obviously, we have told only a small part of the story. This field is so dynamic and rapidly developing that new approaches in modelling and representation of anthropomorphic phantoms will continuously appear. The topic of this book is interdisciplinary and therefore very challenging. To solve a particular problem, knowledge in engineering, physics, biology, mathematics, etc, is required. In this sense, the material in this book may be viewed as a strong base on which solutions to particular problems in medical practice are suggested.

We believe that the new knowledge you obtained will be of benefit for your professional career.

Many thanks!
The Authors

doi:10.1088/2053-2563/aae197ch9

www.ingramcontent.com/pod-product-compliance
Ingram Content Group UK Ltd.
Pitfield, Milton Keynes, MK11 3LW, UK
UKHW051855100926
8556IPUK00005B/630

9 780750 313247